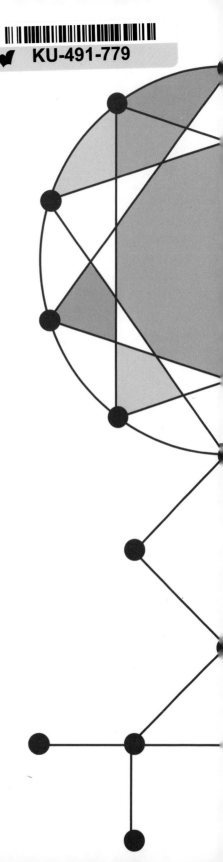

Completely
Positive
Matrices

Completely
Positive
Matrices

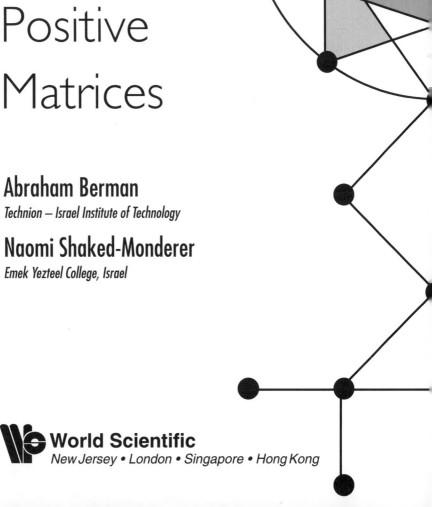

Abraham Berman
Technion – Israel Institute of Technology

Naomi Shaked-Monderer
Emek Yezteel College, Israel

World Scientific
New Jersey • London • Singapore • Hong Kong

Published by

World Scientific Publishing Co. Pte. Ltd.

5 Toh Tuck Link, Singapore 596224

USA office: Suite 202, 1060 Main Street, River Edge, NJ 07661

UK office: 57 Shelton Street, Covent Garden, London WC2H 9HE

British Library Cataloguing-in-Publication Data

A catalogue record for this book is available from the British Library.

1003606592

COMPLETELY POSITIVE MATRICES

ISBN 981-238-368-9

Printed by FuIsland Offset Printing (S) Pte Ltd, Singapore

To our families, who make our lives complete and positive.

Preface

Completely positive matrices are special positive semidefinite matrices. Every real positive semidefinite matrix A can be decomposed as $A = BB^T$ where B is a real matrix. A positive semidefinite matrix A is completely positive if there is such a decomposition where B is entrywise nonnegative. Completely positive matrices should not be confused with the completely positive maps that are studied in operator theory. Completely positive matrices (in the sense of this book) arise, through the related concept of completely positive quadratic forms, in the theory of block designs. Notes on this and on other applications can be found in Section 2.1.

Every matrix of rank r can be written as a sum of r (but not less) rank 1 matrices. If $A = BB^T$, then A can be represented as a sum of the rank 1 matrices $\mathbf{b}_i \mathbf{b}_i^T$, where the $\mathbf{b}_i$'s are the columns of B. If rank $A = r$, it has a decomposition $A = BB^T$ where B has r columns. For this reason, the minimal number of columns in an entrywise nonnegative matrix B such that $A = BB^T$ is called the cp-rank of A.

The two main problems in the theory of completely positive matrices are:

(a) deciding if a given matrix is completely positive.
(b) determining the cp-rank of a given cp matrix.

These two problems are still open. In spite of that, the topic of complete positivity is quite fascinating and quite a lot can be said on the above questions. In this book we try to compose an up-to-date summary of known results.

The book consists of three chapters. The second chapter deals with complete positivity and the third with the cp-rank. The first chapter contains the needed background, including positive semidefinite matrices, nonnegative matrices, Schur complements, cones and graphs. Some of these background topics are covered in greater detail than others, in order to make the book easily accessible even to people with only a basic knowledge of linear algebra. We also expand the discussion when the techniques used can be extended to the study of completely positive matrices.

The book can be used as a reference and as a text for a graduate or advanced undergraduate course or seminar in matrix theory. An almost final version of the book was used in a course on "topics in matrix theory" at the Technion. We are very thankful to the participants of this course for valuable remarks and suggestions.

The core of the book consists of Sections 2.1–2.5 and 3.1–3.3. With an introduction based on the preliminaries chapter it may be used in a one quarter course. Additional sections can be covered in a semesterial course. The dependence relations between the various sections is given in the following diagram:

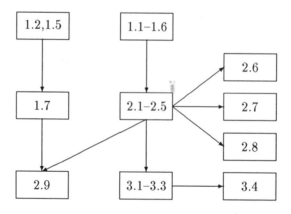

The work on this book was supported by the Fund for Promotion of Research at the Technion. Parts of the book were written during visits of the first author to the Institute for Advanced Studies in Princeton, University of California in San Diego, Technical University of Berlin, University of Padova, University College in Dublin, and Rockefeller University in New York. We are grateful to these fine institutions for their hospitality, and of course to our home institutions, the Technion — Israel Institute of Technology, and Emek Yezreel College.

Haifa, December 2002   Abraham Berman
                                                 Naomi Shaked-Monderer

Contents

Chapter 1

Preliminaries

1.1 Matrix theoretic background

In this section we review matrix theoretical notations, terminology and results. Some of these are very basic, and are mentioned mainly to establish the notations of this book. Some are less basic, but still we do not provide any proofs, and the readers are invited to either try and prove them (see the exercises at the end of the section), or look them up in matrix analysis books (see the notes at the end of the section). Unless stated otherwise, the matrices in this book are real.

$\mathbf{R}^n$ and Euclidean vector spaces

We denote by $\mathbf{R}^n$ the vector space of real column vectors of length n. The entries of $\mathbf{x} \in \mathbf{R}^n$ are denoted by $x_1, \ldots, x_n$. The *nonnegative orthant* of $\mathbf{R}^n$ is denoted by $\mathbf{R}^n_+$, *i.e.*,

$$\mathbf{R}^n_+ = \{\mathbf{x} \in \mathbf{R}^n \,|\, x_i \geq 0 \text{ for every } i = 1, \ldots, n\}.$$

If $\mathbf{x} \in \mathbf{R}^n_+$, we say that $\mathbf{x}$ is a *nonnegative vector* and write $\mathbf{x} \geq 0$. If all the entries of $\mathbf{x}$ are positive, we say that the vector $\mathbf{x}$ is *positive* and write $\mathbf{x} > 0$. For a general vector $\mathbf{x} \in \mathbf{R}^n$, the *support* of $\mathbf{x}$ is

$$\operatorname{supp} \mathbf{x} = \{i \,|\, x_i \neq 0\}.$$

If $\alpha \subseteq \{1, \ldots, n\}$, we denote by $\mathbf{x}[\alpha]$ the vector obtained from $\mathbf{x}$ by erasing every entry x_i, $i \notin \alpha$. For $S \subseteq \mathbf{R}^n$, $\operatorname{Span} S$ denotes the subspace of $\mathbf{R}^n$ spanned by S.

A *Euclidean vector space* is a vector space V over the field of real numbers $\mathbf{R}$, endowed with an inner product $\langle \cdot, \cdot \rangle$. For $V = \mathbf{R}^n$ we denote by

$\langle \cdot, \cdot \rangle$ the *standard inner product* on $\mathbf{R}^n$, and by $\| \cdot \|$ the norm it induces, the 2-*norm*. That is, for $\mathbf{x}, \mathbf{y} \in \mathbf{R}^n$,

$$\langle \mathbf{x}, \mathbf{y} \rangle = \mathbf{x}^T \mathbf{y} = \sum_{i=1}^{n} x_i y_i \, ,$$

$$\|\mathbf{x}\| = \sqrt{\mathbf{x}^T \mathbf{x}} = \left(\sum_{i=1}^{n} x_i^2 \right)^{1/2} .$$

The *angle* between two nonzero vectors $\mathbf{x}, \mathbf{y}$ is the unique $0 \leq \theta \leq \pi$ such that

$$\cos \theta = \frac{\langle \mathbf{x}, \mathbf{y} \rangle}{\|\mathbf{x}\| \|\mathbf{y}\|}.$$

We denote by $\mathbf{e}_i$, $i = 1, \ldots, n$, the vectors of the *standard basis* of $\mathbf{R}^n$. That is, $\mathbf{e}_i$ is the vector all of whose entries are zero, except for 1 in the i-th position. The vector with all entries equal to 1 is denoted by $\mathbf{e}$.

Matrices and determinants

We denote the vector space of $m \times n$ real matrices by $\mathbf{R}^{m \times n}$ ($\mathbf{R}^{n \times 1} = \mathbf{R}^n$). For a matrix A, A^T denotes the transpose of A, cs A denotes the column space of A, rs A denotes the row space of A and ns A denotes the null space of A.

Some special matrices: $0_{m \times n}$ is the zero matrix in $\mathbf{R}^{m \times n}$, $J_{m \times n}$ is the $m \times n$ matrix of all ones; If $m = n$ we also write 0_n and J_n. I_n is the $n \times n$ identity matrix. We omit the subscripts when the orders of the matrices are clear. A square matrix all of whose entries are zero except for 1 in the ij position will be denoted by E_{ij} (its order will always be clear from the context).

A *permutation matrix* P is a square matrix which has exactly one nonzero entry, equal to 1, in each row and each column. Multiplying a matrix A from the left by P results in a permutation of the rows of A (that is, if $p_{1i_1}, p_{2i_2} \ldots, p_{ni_n}$ are the nonzero elements in P, then PA is the matrix whose j-th row is the i_j-th row of A). Multiplying A from the right by P^T has the same effect on the columns of A. Hence PAP^T is a matrix obtained from A by permuting both the rows and the columns of A the same way.

Given a vector $\mathbf{d} \in \mathbf{R}^n$, $\mathbf{d}^T = (d_1, \ldots, d_n)$, we denote by $\mathrm{diag}(\mathbf{d})$ or $\mathrm{diag}(d_1, \ldots, d_n)$ the $n \times n$ diagonal matrix with $d_1, \ldots, d_n$ as diagonal elements. A matrix D is a *positive diagonal matrix* if D is diagonal and all its diagonal elements are positive. If D is a positive diagonal matrix, we say that DAD is obtained from A by a *positive diagonal congruence*.

If A is an $m \times n$ matrix, α and β are sets of indices, $\alpha \subseteq \{1, \ldots, m\}$ and $\beta \subseteq \{1, \ldots, n\}$, then $A[\alpha|\beta]$ is the submatrix of A with rows indexed by the elements of α in increasing order and columns indexed by the elements of β in increasing order. $A[\alpha|\beta)$ is the submatrix of A with rows indexed by α and columns by $\{1, \ldots, n\} \setminus \beta$ (both index sets in increasing order). $A(\alpha|\beta]$ and $A(\alpha|\beta)$ are defined in a similar way. For square matrices, $A[\alpha|\alpha]$ is abbreviated by $A[\alpha]$, and $A(\alpha|\alpha)$ by $A(\alpha)$. We write $A[i_1, i_2, \ldots, i_k]$ instead of $A[\{i_1, i_2, \ldots, i_k\}]$. The matrix $A[\alpha]$ is a *principal submatrix* of A, and if $\alpha = \{1, \ldots, k\}$ for some $1 \leq k \leq n$, then it is a *leading principal submatrix* of A. The determinants of such submatrices are a *principal minor* and a *leading principal minor*, respectively.

An $n \times n$ matrix A is *reducible* if for some $\alpha \subseteq \{1, \ldots, n\}$ the submatrix $A[\alpha|\alpha)$ is equal to 0. Equivalently, A is reducible if for some permutation matrix P, PAP^T has the block form

$$\begin{pmatrix} A & B \\ 0 & C \end{pmatrix},$$

where A and C are square. A square matrix which is not reducible is *irreducible*.

If A and B are square matrices, their *direct sum* $A \oplus B$ is the matrix in block form

$$\begin{pmatrix} A & 0 \\ 0 & B \end{pmatrix}.$$

If A is any matrix, we denote by $|A|$ the matrix whose elements are the absolute values of the corresponding elements in A: $|A|_{ij} = |a_{ij}|$. The *comparison matrix* of a square matrix A is denoted by $M(A)$. It is defined by:

$$M(A)_{ij} = \begin{cases} |a_{ij}|, & \text{if } i = j \\ -|a_{ij}|, & \text{if } i \neq j \end{cases}$$

Another concept that we will need is that of a generalized inverse of a matrix.

Theorem 1.1 *If A is an $m \times n$ matrix, there exists a unique $n \times m$ matrix X such that $AXA = A$, $XAX = X$ and AX and XA are symmetric.*

Definition 1.1   The unique matrix X in Theorem 1.1 is called the *Moore-Penrose generalized inverse* of A, and is denoted by $A^\dagger$.

Of course, if A is invertible, $A^\dagger = A^{-1}$. Here is another example:

Example 1.1 $J_n^\dagger = \frac{1}{n^2} J_n$.

The importance of the Moore-Penrose generalized inverse lies in the fact that $A^\dagger \mathbf{b}$ is the best least squares solution of $A\mathbf{x} = \mathbf{b}$. That is, it is the vector of minimal 2-norm for which the 2-norm of $A\mathbf{x} - \mathbf{b}$ is minimal. In other words, the vector $A^\dagger \mathbf{b}$ is the projection of $\mathbf{b}$ on cs A.

In Section 1.3 we will need the *Cauchy-Binet Formula* for the determinant of the product of two matrices:

Theorem 1.2 [The Cauchy-Binet Formula] *If A is $n \times m$ and B is $m \times n$, $n \le m$, then*

$$\det(AB) = \sum_{\substack{|\alpha|=n \\ \alpha \subseteq \{1,\dots,m\}}} \det A[1,\dots,n|\alpha]B[\alpha|1,\dots,n]. \qquad (1.1)$$

($|\alpha|$ denotes the cardinality of the set α.)

Example 1.2   Let

$$A = \begin{pmatrix} 1 & 2 & 3 \\ 4 & 5 & 6 \end{pmatrix} \text{ and } B = \begin{pmatrix} 7 & 8 \\ 9 & 0 \\ 1 & 2 \end{pmatrix}.$$

Then, by the Cauchy-Binet Formula (1.1),

$$\begin{aligned}
\det(AB) \; = \; & \det \begin{pmatrix} 1 & 2 \\ 4 & 5 \end{pmatrix} \det \begin{pmatrix} 7 & 8 \\ 9 & 0 \end{pmatrix} + \\
& \det \begin{pmatrix} 1 & 3 \\ 4 & 6 \end{pmatrix} \det \begin{pmatrix} 7 & 8 \\ 1 & 2 \end{pmatrix} + \\
& \det \begin{pmatrix} 2 & 3 \\ 5 & 6 \end{pmatrix} \det \begin{pmatrix} 9 & 0 \\ 1 & 2 \end{pmatrix}.
\end{aligned}$$

In Section 1.4 we will use the following useful determinantal identity due to Sylvester:

Theorem 1.3 [Sylvester's Determinantal Identity] *Let A be an $n \times n$ matrix and let α be a subset of cardinality k of $\{1,\dots,n\}$. Arrange the $n-k$*

numbers in $\{1, \ldots, n\} \setminus \alpha$ *in their natural increasing order,* $i_1 < \ldots < i_{n-k}$, *and let*

$$b_{pq} = \det A[\alpha \cup \{i_p\} | \alpha \cup \{i_q\}],$$

$B = (b_{pq})_{p,q=1}^{n-k}$. *Then*

$$\det B = (\det A[\alpha])^{n-k-1} \det A. \tag{1.2}$$

Example 1.3 Let

$$A = \begin{pmatrix} 1 & 2 & 3 \\ 4 & 5 & 6 \\ 7 & 8 & 9 \end{pmatrix}, \quad \alpha = \{2\}.$$

Then

$$b_{11} = \det \begin{pmatrix} 1 & 2 \\ 4 & 5 \end{pmatrix}, \; b_{12} = \det \begin{pmatrix} 2 & 3 \\ 5 & 6 \end{pmatrix}$$

$$b_{21} = \det \begin{pmatrix} 4 & 5 \\ 7 & 8 \end{pmatrix}, \; b_{22} = \det \begin{pmatrix} 5 & 6 \\ 8 & 9 \end{pmatrix}.$$

The minor $\det A[\alpha \cup \{i\} | \alpha \cup \{j\}]$ $(i, j \notin \alpha)$ is called a *bordered minor* of $A[\alpha]$ in A.

We conclude this part with the determinant of a Cauchy matrix. An $n \times n$ *Cauchy matrix* is a matrix C such that

$$c_{ij} = \frac{1}{x_i + y_j}, \quad i, j = 1, \ldots, n, \tag{1.3}$$

where $x_1, \ldots, x_n$ and $y_1, \ldots, y_n$ are two sequences of numbers such that $x_i \neq -y_j$ for every $i, j = 1, \ldots, n$. If $x_i = y_i$, $i = 1, \ldots, n$, C is a *symmetric Cauchy matrix*. A *Hilbert matrix* is a special case of a symmetric Cauchy matrix, where $x_i = y_i = i - 1/2$, $i = 1, \ldots, n$. We denote the $n \times n$ Hilbert matrix by $H(n)$. So the ij element of $H(n)$ is

$$\frac{1}{i + j - 1}.$$

Let C be a Cauchy matrix given by (1.3), then

$$\det C = \frac{\prod_{1 \le j < i \le n}(x_i - x_j) \prod_{1 \le j < i \le n}(y_i - y_j)}{\prod_{i,j=1}^{n}(x_i + y_j)}. \tag{1.4}$$

Eigenvalues and eigenvectors

We denote the *characteristic polynomial* of an $n \times n$ matrix A by $\Delta_A(x)$, *i.e.*, $\Delta_A(x) = \det(xI - A)$. We will need the following formula for the coefficients of $\Delta_A(x)$:

Theorem 1.4 *For any $n \times n$ matrix A,*

$$\Delta_A(x) = x^n - p_1 x^{n-1} + p_2 x^{n-2} - \ldots + (-1)^k p_k x^{n-k} + \ldots + (-1)^n p_n,$$

where p_k is the sum of all $k \times k$ principal minors of A.

Recall that if

$$A\mathbf{x} = \lambda\mathbf{x} \tag{1.5}$$

for some vector $\mathbf{x} \neq \mathbf{0}$, then λ is an *eigenvalue* of A, and $\mathbf{x}$ is an *eigenvector* of A associated with λ. For a real matrix A, λ may be a real or complex number, and $\mathbf{x}$ may be a vector with real or complex entries. The eigenvalues of A are all the roots (real and complex) of the characteristic polynomial $\Delta_A(x)$. The eigenvalue λ is a *simple eigenvalue* of A if the multiplicity of λ as a zero of $\Delta_A(x)$ is 1. Equivalently, λ is a simple eigenvalue of A if the dimension of the *eigenspace* of A corresponding to λ (that is, the subspace consisting of all vectors which satisfy (1.5)) is of dimension 1.

The set of all eigenvalues of A is the *spectrum* of A. The *spectral radius* of A, $\rho(A)$, is the maximum modulus of an eigenvalue of A. Clearly, the closed disc in the complex plane centered at the origin with radius $\rho(A)$ contains all the eigenvalues of A, and it is the smallest such disc. The next result, Geršgorin's Theorem, describes a smaller set that covers the spectrum of $n \times n$ (real or complex) matrix A, consisting of a union of closed discs:

Theorem 1.5 [Geršgorin's Theorem] *If A is an $n \times n$ matrix and*

$$R_i = \sum_{\substack{j=1 \\ j \neq i}}^n a_{ij}, \quad i = 1, \ldots, n,$$

then all the eigenvalues of A are contained in the union of n closed discs D_i, $i = 1, \ldots, n$, where

$$D_i = \{z \mid |z - a_{ii}| \leq R_i\}.$$

Moreover, if A is irreducible and an eigenvalue λ of A lies on the boundary of $\cup_{i=1}^{n} D_i$, then λ lies on the boundary of each of the discs D_i, that is, $|\lambda - a_{ii}| = R_i$ for every $i = 1, \ldots, n$.

Since the spectrum of A is the same as that of A^T, a similar result can be stated using the sums of the off-diagonal elements in each column.

Theorem 1.5 can be used to obtain conditions on the elements of a matrix which guarantee invertibility.

Definition 1.2 A matrix A is *diagonally dominant* if

$$|a_{ii}| \geq \sum_{\substack{j=1 \\ j \neq i}}^{n} |a_{ij}| \text{ for all } i = 1, \ldots, n.$$

It is *strictly diagonally dominant* if

$$|a_{ii}| > \sum_{\substack{j=1 \\ j \neq i}}^{n} |a_{ij}| \text{ for all } i = 1, \ldots, n.$$

It is *irreducibly diagonally dominant* if it is irreducible and diagonally dominant with strict inequality for at least one index i.

Note that there are irreducible diagonally dominant matrices which are not irreducibly diagonally dominant. For example, $(n-2)I_n + J_n$.

Corollary 1.1 *Strictly diagonally dominant matrices and irreducibly diagonally dominant matrices are invertible.*

Since most of the matrices in this book are symmetric, we conclude this section with some special attention to such matrices. First we remind the reader that real symmetric matrices are orthogonally diagonalizable:

Theorem 1.6 *If A is an $n \times n$ real symmetric matrix, then all the eigenvalues of A are real, and there exists an orthogonal $n \times n$ matrix U (that is, $U^T U = I$), such that $U^T A U = D$, where D is a diagonal matrix, with the eigenvalues of A on the diagonal.*

For a real symmetric matrix we denote by $i(A)$ the *inertia* of A. The inertia $i(A)$ of A is a triple $(i_+(A), i_-(A), i_0(A))$, where $i_+(A)$ is the number of positive eigenvalues of A, $i_-(A)$ is the number of negative eigenvalues of A and $i_0(A)$ is the multiplicity of 0 as an eigenvalue of A. Recall that Sylvester's Law of Inertia says that the inertia of an Hermitian matrix is invariant under (Hermitian) congruence. The theorem is stated bellow for real matrices.

Theorem 1.7 [Sylvester's Law of Inertia] *Let A and B be two real symmetric matrices. Then there is a nonsingular real matrix S such that $B = S^T A S$ if and only if A and B have the same inertia.*

The eigenvalues of a general $n \times n$ matrix depend continuously on the entries of the matrix. In general, making this intuitive statement precise is cumbersome, but in the case of real symmetric $n \times n$ matrices it is quite easy. Since the eigenvalues of a real symmetric matrix A are all real, we may put them in increasing order

$$\lambda_1(A) \le \lambda_2(A) \le \ldots \le \lambda_n(A).$$

We can then state:

Theorem 1.8 *If A_t are all symmetric matrices, and $\lim_{t \to t_0} A_t = A$ (entrywise), then for every $i = 1, \ldots, n$,*

$$\lambda_i(A) = \lim_{t \to t_0} \lambda_i(A_t).$$

We conclude this section with an interlacing theorem, describing a relation between the eigenvalues of a real symmetric matrix A and those of a principal submatrix of A.

Theorem 1.9 *Let A by a real symmetric matrix, and $B = A[1, \ldots, n]$. Then*

$$\lambda_1(A) \le \lambda_1(B) \le \lambda_2(A) \le \lambda_2(B) \le \ldots \le \lambda_{n-1}(B) \le \lambda_n(A).$$

Exercises

1.1   Let A and B be as in Theorem 1.2. Use Theorem 1.4 and the fact that $\Delta_{BA}(x) = x^{n-m} \Delta_{AB}(x)$ to prove the Cauchy-Binet Formula (1.1).

1.2   Prove formula (1.4) for the determinant of a Cauchy matrix.
Hint: You can start by subtracting the last column from all other columns and then dividing each row and each column by its common factor.

1.3   Calculate the determinant of $H(n)$, the $n \times n$ Hilbert matrix.

1.4   Use Theorem 1.5 to prove Corollary 1.1.

1.5   Show that

(a) If $D = \text{diag}(d_1, \ldots, d_n)$ is a diagonal matrix, then $D^\dagger$ is the diagonal matrix such that

$$(D^\dagger)_{ii} = \begin{cases} 1/d_i, & \text{if } d_i \neq 0, \\ 0, & \text{if } d_i = 0 \end{cases}.$$

(b) If A is a real symmetric matrix, $A = UDU^T$ where D is a diagonal matrix and U is an orthogonal matrix, then $A^\dagger = UD^\dagger U^T$.

1.6 Prove the following properties of the Moore-Penrose generalized inverse:

(a) $(A^\dagger)^\dagger = A$
(b) $(A^\dagger)^T = (A^T)^\dagger$
(c) $A^T A A^\dagger = A^T = A^\dagger A A^T$
(d) $(AA^T)^\dagger = (A^T)^\dagger A^\dagger$.
(e) For every $\mathbf{x} \in \text{cs}\, A$, $AA^\dagger \mathbf{x} = \mathbf{x}$.

Notes. General references on Matrix Theory are [Gantmacher (1959)], [Horn and Johnson (1985)], [Horn and Johnson (1991)] and [Marcus and Minc (1964)]. A classical reference on generalized inverses is [Ben-Israel and Greville (2003)]. A beautiful expository article on Hilbert matrices is [Choi (1983)]. The Hilbert matrix $H(n)$ has many lovely features. For example, all the entries of its inverse are integers, *e.g.* [Choi (1983)], [Berman and Gueron (2002)]. It is highly ill conditioned and as such useful in testing numerical algorithms. As we will see later, the Hilbert matrices are totally positive and completely positive.

In the discussion of diagonal dominance we follow [Varga (2000)]. Theorem 1.5 is a combination of the original Geršgorin's Theorem and an improvement due to Taussky.

1.2 Positive semidefinite matrices

In this section we review the definition and properties of positive semidefinite matrices. This is not a comprehensive survey — we restrict our attention to real positive semidefinite matrices, and to properties that will be needed in our discussion of completely positive matrices.

Definition 1.3 A matrix $A \in \mathbf{R}^{n \times n}$ is *positive semidefinite* if it is symmetric and $\mathbf{x}^T A \mathbf{x}$ is nonnegative for every $\mathbf{x} \in \mathbf{R}^n$.

We use $A \succeq 0$ to denote that A is positive semidefinite, and write $A \succeq B$ if $A - B \succeq 0$. Note that an identity matrix is positive semidefinite, and so is a square zero matrix. Another simple example is

$$A = \begin{pmatrix} 1 & -1 \\ -1 & 1 \end{pmatrix},$$

for which $\mathbf{x}^T A \mathbf{x} = (x_1 - x_2)^2$.

Proposition 1.1 *The sum of positive semidefinite matrices is positive semidefinite. Also, if A is a positive semidefinite matrix and $a > 0$, then aA is also positive semidefinite.*

Proof. The first assertion follows from the equality $\mathbf{x}^T (A+B)\mathbf{x} = \mathbf{x}^T A \mathbf{x} + \mathbf{x}^T B \mathbf{x}$, and the second from the equality $\mathbf{x}^T (aA)\mathbf{x} = a\mathbf{x}^T A \mathbf{x}$. □

The following proposition is also easy to prove:

Proposition 1.2 *If A is an $n \times n$ positive semidefinite matrix, and S is any $n \times m$ matrix, then $S^T AS$ is positive semidefinite.*

Proposition 1.3 *Any principal submatrix of a positive semidefinite matrix is also positive semidefinite.*

Proof. Let A be an $n \times n$ positive semidefinite matrix, and let $A[\alpha]$ be a principal submatrix of A, $\alpha \subseteq \{1, 2, \ldots, n\}$. Consider $\mathbf{y}^T A[\alpha]\mathbf{y}$. Let $\mathbf{x}$ be a vector of n entries such that $\mathbf{y} = \mathbf{x}[\alpha]$ and $x_i = 0$ for $i \notin \alpha$. Then $\mathbf{y}^T A[\alpha]\mathbf{y} = \mathbf{x}^T A \mathbf{x} \geq 0$. □

Being symmetric, any $n \times n$ real positive semidefinite matrix A has n real eigenvalues (counting multiplicities), and is orthogonally diagonalizable. That is, there exists an orthogonal matrix U such that $U^T AU = D$ is diagonal. In particular,

$$\text{rank } A = \text{rank } D = \text{the number of nonzero eigenvalues of } A.$$

Using these facts (and some others) we derive several characterizations of positive semidefinite matrices. These are stated in Theorem 1.10. For one of them we need the notion of a Gram matrix:

Definition 1.4 Let $\mathbf{v}_1, \ldots, \mathbf{v}_n$ be vectors in an Euclidean space V. The matrix of inner products

$$A = \begin{pmatrix} \langle \mathbf{v}_1, \mathbf{v}_1 \rangle & \langle \mathbf{v}_1, \mathbf{v}_2 \rangle & \cdots & \langle \mathbf{v}_1, \mathbf{v}_n \rangle \\ \langle \mathbf{v}_2, \mathbf{v}_1 \rangle & \langle \mathbf{v}_2, \mathbf{v}_2 \rangle & \cdots & \langle \mathbf{v}_2, \mathbf{v}_n \rangle \\ \vdots & \vdots & & \vdots \\ \langle \mathbf{v}_n, \mathbf{v}_1 \rangle & \langle \mathbf{v}_n, \mathbf{v}_2 \rangle & \cdots & \langle \mathbf{v}_n, \mathbf{v}_n \rangle \end{pmatrix}$$

is called the *Gram matrix* of $\mathbf{v}_1, \ldots, \mathbf{v}_n$, and denoted $\mathrm{Gram}(\mathbf{v}_1, \ldots, \mathbf{v}_n)$.

We can now state the following characterizations.

Theorem 1.10 *Let A be an $n \times n$ real symmetric matrix. Then the following statements are equivalent:*

(a) *A is positive semidefinite.*
(b) *All the eigenvalues of A are nonnegative.*
(c) *All the principal minors of A are nonnegative.*
(d) *There exists an $n \times k$ real matrix B such that $A = BB^T$.*
(e) *There exists an $n \times n$ lower triangular real matrix L such that $A = LL^T$.*
(f) *There exists an $n \times n$ real symmetric matrix C such that $A = C^2$.*
(g) *There exist a k-dimensional Euclidean vector space V and vectors $\mathbf{v}_1, \ldots, \mathbf{v}_n \in V$ such that $A = \mathrm{Gram}(\mathbf{v}_1, \ldots, \mathbf{v}_n)$.*
(h) *There exist k vectors $\mathbf{b}_1, \ldots, \mathbf{b}_k \in \mathbf{R}^n$ such that $A = \sum_{i=1}^{k} \mathbf{b}_i \mathbf{b}_i^T$.*

Proof. We prove the following implications: (a)$\Rightarrow$(b), (a)$\Rightarrow$(c)$\Rightarrow$(b)$\Rightarrow$(f)$\Rightarrow$(d)$\Rightarrow$(g)$\Rightarrow$(a), (d)$\Rightarrow$(e)$\Rightarrow$(d), and (b)$\Leftrightarrow$(h).

(a)$\Rightarrow$(b): Let $A\mathbf{x} = \lambda\mathbf{x}$, $\mathbf{x} \neq \mathbf{0}$. Then $\mathbf{x}^T A \mathbf{x} = \lambda \mathbf{x}^T \mathbf{x} \geq 0$ and since $\mathbf{x}^T\mathbf{x} > 0$, $\lambda = \mathbf{x}^T A \mathbf{x} / \mathbf{x}^T \mathbf{x} \geq 0$.

(a)$\Rightarrow$(c): Let $A[\alpha]$ be a principal submatrix of A. By Proposition 1.3, $A[\alpha]$ is also positive semidefinite, and by the implication (a)$\Rightarrow$(b), all its eigenvalues are nonnegative. Since the determinant is the product of the eigenvalues, $\det A[\alpha]$ is nonnegative.

(c)$\Rightarrow$(b): Let

$$\Delta_A(x) = x^n - p_1 x^{n-1} + p_2 x^{n-2} - \ldots + (-1)^k p_k x^{n-k} + \ldots + (-1)^n p_n$$

be the characteristic polynomial of A. Then p_k is the sum of all $k \times k$ principal minors of A, and since (c) holds, $p_k \geq 0$, $k = 1, 2, \ldots, n$. Suppose $x < 0$. If n is even, then x^n is positive and all other terms in $\Delta_A(x)$

are nonnegative. If n is odd, then x^n is negative and all other terms in $\Delta_A(x)$ are nonpositive. This shows that A cannot have negative eigenvalues. Being symmetric, all the eigenvalues of A are real, which means that the eigenvalues are nonnegative.

(b)$\Rightarrow$(f): Since A is symmetric with nonnegative eigenvalues, it is orthogonally similar to a nonnegative diagonal matrix D. That is, $A = UDU^T$ for some orthogonal matrix U and a nonnegative diagonal matrix $D = \mathrm{diag}(d_1, \ldots, d_n)$. But then $A = U\sqrt{D}U^T U\sqrt{D}U^T$, where $\sqrt{D} = \mathrm{diag}(\sqrt{d_1}, \ldots, \sqrt{d_n})$. Hence $A = C^2$ where $C = U\sqrt{D}U^T$.

(d)$\Rightarrow$(e): To prove this implication we use the fact that any real matrix C has a QR factorization, *i.e.*, $C = QR$ where the rows of Q are orthonormal and R is upper triangular. Let $A = BB^T$, and let $B^T = QR$ be a QR factorization of B^T. Then $A = R^T Q^T QR = LL^T$, where $L = R^T$ is lower triangular. (see also Exercise 1.29 for another proof of this implication.)

(d)$\Rightarrow$(g): Let $A = BB^T$, where B is an $n \times k$ matrix. Let $V = \mathbf{R}^k$, and let $\mathbf{v}_i^T$ be the i-th row of B. Then $A = \mathrm{Gram}(\mathbf{v}_1, \ldots, \mathbf{v}_n)$.

(g)$\Rightarrow$(a): Let $A = \mathrm{Gram}(\mathbf{v}_1, \ldots, \mathbf{v}_n)$, and let $\mathbf{x} \in \mathbf{R}^n$. Then

$$
\begin{aligned}
\mathbf{x}^T A\mathbf{x} = \sum_{i,j=1}^{n} a_{ij}x_i x_j &= \sum_{i,j=1}^{n} \langle \mathbf{v}_i, \mathbf{v}_j \rangle x_i x_j \\
&= \sum_{i,j=1}^{n} \langle x_i\mathbf{v}_i, x_j\mathbf{v}_j \rangle \\
&= \left\langle \sum_{i=1}^{n} x_i\mathbf{v}_i, \sum_{j=1}^{n} x_j\mathbf{v}_j \right\rangle = \left\| \sum_{i=1}^{n} x_i\mathbf{v}_i \right\|^2 \geq 0.
\end{aligned}
$$

(b)$\Leftrightarrow$(h): $A = BB^T$ iff $A = \sum_{i=1}^{k} \mathbf{b}_i \mathbf{b}_i^T$, where $\mathbf{b}_i$, $i = 1, \ldots, k$, are the columns of B.

The implications (e)$\Rightarrow$(d) and (f)$\Rightarrow$(d) are clear, so the proof is complete. $\square$

Using the last theorem we may obtain many examples of positive semidefinite matrices. For example, $J = ee^T$ is positive semidefinite.

Remark 1.1   By Theorem 1.10(b), the matrix C constructed in our proof of (b)$\Rightarrow$(f) was actually positive semidefinite. It can be shown that for any positive semidefinite matrix A there is a *unique* positive semidefinite matrix C satisfying $A = C^2$. It is called the *square root* of A, and denoted by $\sqrt{A}$. The matrix B in (d), however, is not unique. In particular, if A is a rank

r positive semidefinite matrix, there exists an $n \times r$ matrix B such that $A = BB^T$. To see that, observe that there exist a unitary matrix U and a diagonal matrix $D = \text{diag}(d_1, \ldots, d_n)$ such that $U^T A U = D$. The d_i's are the eigenvalues of A. They are nonnegative and exactly r of them are positive, say $d_1, d_2, \ldots, d_r > 0$. Thus $A = \hat{B}\hat{B}^T$, where $\hat{B} = U\sqrt{D}$. Since $\hat{B}[1, \ldots, n | r + 1, \ldots, n] = 0$, $A = BB^T$, where $B = \hat{B}[1, \ldots, n | 1, \ldots, r]$.

We leave the proof of the next proposition as an exercise:

Proposition 1.4 *If $A = BB^T$, then $\text{ns}\, A = \text{ns}\, B^T$ and $\text{cs}\, A = \text{cs}\, B$.*

If a diagonal entry in a positive semidefinite matrix is equal to zero, then all entries in the corresponding row and column are zeros. This observation follows from the fact that for every i and j,

$$\det \begin{pmatrix} a_{ii} & a_{ij} \\ a_{ij} & a_{jj} \end{pmatrix} \geq 0.$$

It is a special case of the following Principal Submatrix Rank Property, whose proof is also left as an exercise.

Proposition 1.5 *Let $A[\alpha]$ be a principal submatrix of an $n \times n$ positive semidefinite matrix A, let $A[\alpha | 1, 2, \ldots, n]$ be the submatrix of A based on the rows indexed by α, and let $A[1, 2, \ldots, n | \alpha]$ be the submatrix of A based on the columns indexed by α. Then*

$$\text{rank}\, A[\alpha | 1, \ldots, n] = \text{rank}\, A[1, \ldots, n | \alpha] = \text{rank}\, A[\alpha].$$

Using Theorem 1.10 we deduce some additional results.

Proposition 1.6 *If A is positive semidefinite and k is a positive integer, then A^k is positive semidefinite.*

Proof. By Theorem 1.10(f), $A = C^2$, where C is symmetric. Therefore $A^k = (C^2)^k = (C^k)^2$ is positive semidefinite. $\qquad\square$

The next corollary follows from Propositions 1.6 and 1.1:

Corollary 1.2 *If $f(x) = \sum_{i=0}^{m} a_i x^i$ is a polynomial with nonnegative coefficients, and A is positive semidefinite, then $f(A) = \sum_{i=0}^{m} a_i A^i$ is also positive semidefinite.*

We also consider two other types of products of matrices:

Definition 1.5 The *Hadamard product* of $A, B \in \mathbf{R}^{m \times n}$ is the matrix $C = A \circ B$, defined by the entry by entry product. That is, $c_{ij} = a_{ij} b_{ij}$.

Definition 1.6   The *Kronecker product* of $A \in \mathbf{R}^{m \times n}$ and $B \in \mathbf{R}^{p \times q}$ is the block matrix $A \otimes B$ whose ij block is $a_{ij}B$.

Proposition 1.7   *The Hadamard product of positive semidefinite matrices is positive semidefinite.*

Proof.   Let $A = \sum_{i=1}^{k} \mathbf{v}_i \mathbf{v}_i^T$ and $B = \sum_{j=1}^{l} \mathbf{w}_j \mathbf{w}_j^T$. Then $A \circ B = \sum_{i=1,j=1}^{k,l} (\mathbf{v}_i \circ \mathbf{w}_j)(\mathbf{v}_i \circ \mathbf{w}_j)^T$.   □

Definition 1.7   Let k be a positive integer. The *Hadamard k-th power* of a matrix A, $A^{(k)}$, is the Hadamard product of k copies of A:

$$A^{(k)} = A \circ A \circ \ldots \circ A.$$

Corollary 1.3   *If A is positive semidefinite and k is a positive integer, then $A^{(k)}$ is also positive semidefinite.*

To prove similar results for the Kronecker product we need to use two basic properties of this product. These properties are stated in the next lemma, and we leave the proof as an exercise.

Lemma 1.1

(a) *If AC and BD are defined, then $(A \otimes B)(C \otimes D) = AC \otimes BD$.*
(b) $(A \otimes B)^T = A^T \otimes B^T$.

Corollary 1.4   *The Kronecker product of positive semidefinite matrices is positive semidefinite.*

Proof.   If $A = BB^T$ and $C = FF^T$ then, by Lemma 1.1,

$$A \otimes C = BB^T \otimes FF^T = (B \otimes F)(B^T \otimes F^T) = (B \otimes F)(B \otimes F)^T.$$ □

Remark 1.2   Proposition 1.7 follows from the last corollary, since $A \circ C$ is a principal submatrix of $A \otimes C$.

Positive semidefinite matrices which are diagonally dominant will play a role in our discussion of completely positive matrices. Recall that a matrix $A \in \mathbf{R}^{n \times n}$ is diagonally dominant if for every $i = 1, \ldots, n$

$$|a_{ii}| \geq \sum_{\substack{j=1 \\ j \neq i}}^{n} |a_{ij}|.$$

Geršgorin's Theorem (Theorem 1.5) implies that the eigenvalues of a diagonally dominant matrix lie in the right halfplane $\{z \in \mathbf{C} \mid \operatorname{Re} z \geq 0\}$. We state here this result in the case that A is a real symmetric matrix, and provide an alternative proof for this case.

Proposition 1.8   *A diagonally dominant symmetric matrix with nonnegative diagonal entries is positive semidefinite.*

Moreover, if A is irreducible, and

$$a_{ii} > \sum_{\substack{j=1 \\ j \neq i}}^{n} |a_{ij}|$$

for at least one i, then A is nonsingular.

Proof.   We prove the first part. Let A be an $n \times n$ symmetric diagonally dominant matrix with nonnegative diagonal entries. Let A_{ij} be the matrix whose principal submatrix based on rows and columns i, j is

$$\begin{pmatrix} |a_{ij}| & a_{ij} \\ a_{ij} & |a_{ij}| \end{pmatrix},$$

and all its other entries are equal to zero. Then each A_{ij} is a positive semidefinite matrix (of rank at most 1), and $A - \sum_{i,j=1}^{n} A_{ij}$ is a nonnegative diagonal matrix. The matrix A is therefore positive semidefinite, as a sum of positive semidefinite matrices. $\square$

We conclude the section with a discussion of positive definite matrices.

Definition 1.8   A matrix $A \in \mathbf{R}^{n \times n}$ is *positive definite* if it is symmetric and $\mathbf{x}^T A \mathbf{x} > 0$ for every $0 \neq \mathbf{x} \in \mathbf{R}^n$.

Proposition 1.9   *An $n \times n$ matrix A is positive definite if and only if it is a nonsingular positive semidefinite matrix.*

Proof.   The "only if " part is obvious. For the "if" part, write the positive semidefinite matrix A as $A = BB^T$. Then $\mathbf{x}^T A \mathbf{x} = \|B^T \mathbf{x}\|^2$. By Proposition 1.4, this implies that $\mathbf{x}^T A \mathbf{x} = 0$ iff $\mathbf{x} \in \operatorname{ns} A$. Hence if A is nonsingular, $\mathbf{x}^T A \mathbf{x} > 0$ for every $0 \neq \mathbf{x} \in \mathbf{R}^n$. $\square$

The following theorem is analogous to Theorem 1.10, but has an additional statement concerning nested minors. We define the notion of nested minors and state the theorem without a proof.

Definition 1.9   Let α_i, $i = 1, \ldots, k$, be subsets of $\{1, \ldots, n\}$ such that for every i, $|\alpha_i| = i$ and $\alpha_i \subseteq \alpha_{i+1}$. Then the minors $\det(A[\alpha_i])$ are *nested principal minors* of A.

The leading minors of A, $\det(A[1, \ldots, i])$, $i = 1, \ldots, n$, are n nested principal minors of A.

Theorem 1.11   *The following statements on an $n \times n$ symmetric matrix A are equivalent:*

(a) *A is positive definite.*
(b) *All the eigenvalues of A are positive.*
(c) *All the principal minors of A are positive.*
(d) *$A = BB^T$ for some nonsingular matrix B.*
(e) *$A = LL^T$, where L is a nonsingular lower triangular matrix.*
(f) *$A = C^2$, where C is a nonsingular symmetric matrix.*
(g) *A is the Gram matrix of n linearly independent vectors.*
(h) *$A = \sum_{i=1}^{n} \mathbf{b}_i \mathbf{b}_i^T$, where $\mathbf{b}_1, \ldots, \mathbf{b}_n \in \mathbf{R}^n$ are linearly independent.*
(i) *A has a set of n positive nested principal minors.*

Example 1.4   The $n \times n$ Hilbert matrix $H(n)$, whose ij entry is $1/(i+j-1)$, is positive definite. To see that, consider $V = C[0,1]$, the space of real continuous functions on $[0, 1]$, with the inner product

$$\langle f, g \rangle = \int_0^1 f(x)g(x)\, dx.$$

Then $H(n)$ is the Gram matrix of $f_i(x) = x^{i-1}$, $i = 1 \ldots, n$.

Example 1.5   The matrix

$$\begin{pmatrix} 0 & 0 \\ 0 & -1 \end{pmatrix}$$

shows that an analogous statement to (i)$\Rightarrow$(a) does not hold in the positive semidefinite case. That is, a matrix with a set of nonnegative nested principal minors is not necessarily positive semidefinite.

In the positive semidefinite case we have the following weaker result:

Proposition 1.10   *Let A be an $n \times n$ symmetric matrix. Let $A[\alpha_i]$, $i = 1, \ldots, n$, be nested minors of A, such that $A[\alpha_i]$, $i = 1, 2, \ldots, n-1$, are positive, and $\det A \geq 0$. Then A is positive semidefinite.*

Note that if A is a positive semidefinite matrix, then for every $\varepsilon > 0$, $A + \varepsilon I$ is positive definite. Hence every positive semidefinite matrix is a limit of positive definite matrices. The converse is also true: any matrix which is a limit of positive definite (or positive semidefinite) matrices is positive semidefinite.

Exercises

1.7   Prove Proposition 1.4.

1.8   Show that if A and B are positive semidefinite and $C = A + B$, then $\operatorname{ns} C = \operatorname{ns} A \cap \operatorname{ns} B$.

1.9   Prove Proposition 1.5.
Hint: Suppose that $\alpha = \{1, 2, \ldots, k\}$. Decompose A as BB^T, where B is a 2×1 block matrix

$$\begin{pmatrix} X \\ Y \end{pmatrix},$$

and X is a $k \times k$ matrix.

1.10   Prove Lemma 1.1.

1.11   Prove Proposition 1.10.
Hint: Assume $\alpha_i = \{1, \ldots, i\}$. For every $\varepsilon > 0$, $A_\varepsilon = A + \varepsilon E_{nn}$ is positive definite.

1.12   Show that if A is symmetric, then

$$\exp(A) = \lim_{n \to \infty} \sum_{k=1}^{n} \frac{A^k}{k!}$$

is positive definite.

1.13   Let a_i, $i = 1, \ldots, n$, be positive numbers, $a_i \neq a_j$ for $i \neq j$. Show that the symmetric Cauchy matrix C whose ij entry is $1/(a_i + a_j)$ is positive definite.

1.14   Prove: A matrix A is positive definite if and only if A^{-1} is positive definite.

1.15   Show that if A is positive definite, then the classical adjoint of A, $\operatorname{adj} A$, is positive definite. Give an example of a symmetric matrix A such that $\operatorname{adj} A$ is positive definite but A is not.

1.16   If A is positive definite, then A can be represented as a polynomial in A^2.

1.17   Show that if A is a positive semidefinite matrix, then

(a) The generalized inverse $A^\dagger$ is also positive semidefinite.

(b) $\sqrt{A^\dagger} = \sqrt{A}^\dagger$.

Hint: See Exercise 1.5.

1.18   Show that if A is a positive semidefinite matrix and $\mathbf{v} \in \mathrm{cs}\, A$, then there exists $\delta > 0$ such that $A - \delta \mathbf{v}\mathbf{v}^T$ is positive semidefinite.
Hint: $\mathbf{v} = \sqrt{A}\mathbf{u}$, and $A - \delta \mathbf{v}\mathbf{v}^T = \sqrt{A}(I - \delta \mathbf{u}\mathbf{u}^T)\sqrt{A}$. Hence it suffices to consider the case $A = I$.

Notes.   As mentioned in the beginning of this section, this is not a comprehensive survey. A good reference on positive semidefinite matrices (real and complex) is [Horn and Johnson (1985)]. Most of Theorem 1.9 follows easily from Theorem 1.10 and Propositions 1.9 and 1.4. The implication (i)⇒(b), however, requires additional theory, such as interlacing (Theorem 1.9). The Hadamard product is also called *Schur product*, since Proposition 1.7 was proved by I. Schur. Apparently the entrywise product of matrices was first used by Moutard in 1894. This fact and other historical remarks on this product can be found in [Horn and Johnson (1991)], which contains an excellent extensive reference on Hadamard products and on Kronecker products.

1.3   Nonnegative matrices and M-matrices

In this section we present some of the main results on nonnegative and related matrices. We include only results which are relevant to the subject of this book, and mostly without proof. The section is divided into three parts: nonnegative matrices, totally nonnegative matrices, and M-matrices.

Nonnegative matrices

Definition 1.10   A matrix A is *nonnegative* if all its entries are nonnegative. A matrix A is *positive* if all its entries are positive.

We write $A \geq 0$ to denote that A is nonnegative, and $A \geq B$ if $A - B$ is nonnegative.

Theorem 1.12 *If A is a nonnegative square matrix, then $\rho(A)$, the spectral radius of A, is an eigenvalue of A, and A has a nonnegative eigenvector corresponding to $\rho(A)$.*

When A is irreducible, more can be said:

Theorem 1.13 [The Perron-Frobenius Theorem] *Let A be a square irreducible nonnegative matrix. Then*

(a) $\rho(A)$ *is a positive simple eigenvalue.*

(b) A *has a positive eigenvector corresponding to $\rho(A)$.*

(c) *If $\lambda_0 = \rho(A), \lambda_1 = \rho(A)e^{i\theta_1}, \ldots, \lambda_{k-1} = \rho(A)e^{i\theta_{k-1}}$ are the eigenvalues of A with absolute value $\rho(A)$, then $\lambda_1, \ldots, \lambda_{k-1}$ are also simple, $\theta_i = 2\pi i/k$, $1 = 1, \ldots, k-1$, and A is similar to $e^{2\pi i/k}A$. In particular, the whole spectrum of A is invariant under a rotation of the complex plane by $2\pi/k$.*

(d) *If k in (c) is greater than 1, then there exists a permutation matrix P such that*

$$P^T AP = \begin{pmatrix} 0 & A_{12} & 0 & \ldots & 0 \\ 0 & 0 & A_{23} & \ldots & 0 \\ \vdots & \vdots & \ddots & \ddots & \vdots \\ 0 & & \ldots & 0 & A_{k-1\,k} \\ A_{k1} & 0 & & \ldots & 0 \end{pmatrix} \qquad (1.6)$$

where the zero blocks on the diagonal are square.

Definition 1.11 The eigenvalue $\rho(A)$ of an irreducible nonnegative matrix A is called the *Perron root* of A, and a positive eigenvector $\mathbf{x}$ such that $A\mathbf{x} = \rho(A)\mathbf{x}$ is called a *Perron vector* of A. The number of eigenvalues of an irreducible nonnegative matrix A of modulus $\rho(A)$ is called the *index of cyclicity* of the irreducible matrix A.

Example 1.6 The matrix

$$A = \begin{pmatrix} 0 & 1 & 0 & 0 \\ 0 & 0 & 1 & 0 \\ 0 & 0 & 0 & 1 \\ 4 & 0 & 3 & 0 \end{pmatrix}$$

is irreducible. Its spectral radius is 2, the eigenvalues are $2, -2, i, -i$. $(1, 2, 4, 8)^T$ is a Perron vector of A. The index of cyclicity is 2. Switching the second and third rows and columns of A yields the matrix

$$\begin{pmatrix} 0 & 0 & 1 & 0 \\ 0 & 0 & 0 & 1 \\ 0 & 1 & 0 & 0 \\ 4 & 3 & 0 & 0 \end{pmatrix}$$

which has the form (1.6).

In view of Theorem 1.13, the index of cyclicity of any square positive matrix is 1:

Corollary 1.5 *If A is a positive square matrix, then $\rho(A)$ is a simple eigenvalue of A, which is greater than the absolute value of any other eigenvalue, and A has a positive eigenvector corresponding to $\rho(A)$.*

Totally nonnegative matrices

Totally nonnegative matrices are an important class of nonnegative matrices.

Definition 1.12   A square matrix is *totally nonnegative* if all its minors are nonnegative, and *totally positive* if all its minors are positive.

Every zero matrix is totally nonnegative, as is every J_n matrix and every identity matrix. A nontrivial 3×3 example is

$$\begin{pmatrix} 3 & 4 & 1 \\ 2 & 3 & 1 \\ 1 & 2 & 2 \end{pmatrix}.$$

Another example:

Example 1.7   A Cauchy matrix

$$C = \left(\frac{1}{x_i + y_j} \right)^n_{i,j=1}$$

is totally positive when both the sequences $\{x_i\}_{i=1}^n$ and $\{y_i\}_{i=1}^n$ are positive and strictly increasing. To see that, observe that every submatrix of C is also a Cauchy matrix, and use the formula (1.4) for the determinant of a Cauchy matrix.

Remark 1.3   Note that a symmetric totally nonnegative matrix is both nonnegative and positive semidefinite. A matrix which has both these properties is called *doubly nonnegative*.

While I_n and J_n are totally nonnegative, $J_n + I_n$ is not (for $n \geq 3$). So the sum of totally nonnegative matrices is not necessarily totally nonnegative. However, by the Cauchy-Binet Formula (1.1), the product of totally nonnegative matrices is totally nonnegative.

Proposition 1.11 *If A and B are $n \times n$ totally nonnegative matrices, then AB is also totally nonnegative. If A and B are totally positive, so is AB.*

Any totally nonnegative matrix may be factored into a product of totally nonnegative matrices of a special form:

Definition 1.13 An $n \times n$ upper triangular matrix U is called *triangular totally positive* if for every $k = 1, \ldots, n$ and sets of k indices, $\alpha = \{i_1, i_2, \ldots, i_k\}$ and $\beta = \{j_1, j_2, \ldots, j_k\}$, ordered increasingly and satisfying $1 \leq i_s \leq j_s \leq n$ for $s = 1, \ldots, k$, the minor $\det A[\alpha|\beta]$ is positive.

An $n \times n$ lower triangular matrix L is called *triangular totally positive* if L^T is triangular totally positive.

A *triangular totally nonnegative* matrix is exactly what the name says — a totally nonnegative matrix which is triangular.

The rest of the minors of a triangular totally positive matrix are necessarily zero by the triangular structure of the matrix, hence a triangular totally positive matrix is totally nonnegative, but not totally positive.

Theorem 1.14 *For every totally nonnegative matrix A there exist a lower triangular totally nonnegative matrix L and an upper triangular totally nonnegative matrix U such that $A = LU$.*

If A is totally positive then L and U are both triangular totally positive.

M-matrices

We consider now square matrices whose off-diagonal entries are nonpositive.

Definition 1.14 A square matrix $A = sI - B$, $B \geq 0$, is an *M-matrix* if $s \geq \rho(B)$.

An M-matrix $A = sI - B$, $B \geq 0$, is nonsingular if and only if $s > \rho(B)$. There are many characterizations of nonsingular M-matrices. Some of them are given in the following theorem.

Theorem 1.15 *Let $A = sI - B$ be an $n \times n$ matrix, $B \geq 0$. Then the following statements are equivalent:*

(a) *A is a nonsingular M-matrix.*

(b) *All the principal minors of A are positive.*

(c) *A is positive stable, i.e., the real parts of the eigenvalues of A are positive.*

(d) *For every permutation matrix P, there exist a lower triangular matrix L with positive diagonal entries and an upper triangular matrix U with positive diagonal entries such that $P^T AP = LU$.*

(e) *There exists a positive diagonal matrix $D = \mathrm{diag}(d_1, \ldots, d_n)$ such that AD is strictly diagonally dominant, i.e., $d_i a_{ii} > \sum_{j \neq i} -d_j a_{ij}$.*

(f) *For every permutation matrix P the leading principal minors of $P^T AP$ are positive.*

(g) *A^{-1} is nonnegative.*

To obtain analogies for some of these characterizations for general (not necessarily nonsingular) M-matrices (see Exercise 1.23), one can use the following observation:

Proposition 1.12   *Let $A = sI - B$, $B \geq 0$. Then A is an M-matrix if and only if $A + \varepsilon I$ is a nonsingular M-matrix for every $\varepsilon > 0$.*

Corollary 1.6   *A matrix $A = sI - B$, where B is a symmetric nonnegative matrix, is an M-matrix if and only if it is positive semidefinite.*

The only theorem that we prove in this section also deals with symmetric M-matrices. The theorem is needed in Section 2.4.

Theorem 1.16   *If A is a symmetric M-matrix, then there exists a positive diagonal matrix D such that DAD is diagonally dominant.*

Proof.   If A is not irreducible, then it is a direct sum of irreducible symmetric M-matrices. So it suffices to prove the theorem in the case that A is irreducible. In that case, Let ρ be the Perron root of B, and let $\mathbf{x}$ be a Perron vector of B. Let $D = \mathrm{diag}(\mathbf{x})$. Then AD is diagonally dominant, since

$$ADe = s\mathbf{x} - \rho\mathbf{x} \geq 0.$$

Because D is a positive diagonal matrix, this implies that $DADe \geq 0$. That is, DAD is a symmetric diagonally dominant matrix.   □

Remark 1.4   By the proof of Theorem 1.16, if A is nonsingular, then DAD is strictly diagonally dominant, and if A is singular, then the sum of each row in DAD is zero.

Exercises

1.19 A square nonnegative matrix A is called *primitive* if A^m is positive for some positive integer m. Show that the index of cyclicity of a primitive matrix is 1. (The converse is also true: an irreducible nonnegative matrix with index of cyclicity 1 is primitive.)

1.20 Prove Proposition 1.11.

1.21 The symmetric Pascal matrix $P(n)$ is defined by $p_{1i} = p_{i1} = 1$ for every $1 \leq i \leq n$, and $p_{ij} = p_{i-1\,j} + p_{i\,j-1}$ for $2 \leq i, j \leq n$. Prove that $P(n)$ is totally positive.
Hint: Find a nonsingular triangular totally nonnegative matrix S such that $P(n+1) = S(1 \oplus P(n))S^T$.

1.22 Suppose that $A = sI - B$, $B \geq 0$ is irreducible. Prove that A is a nonsingular M-matrix if and only if A^{-1} is positive.

1.23 Prove: Let $A = sI - B$, $B \geq 0$. Then the following statements are equivalent.

(a) A is an M-matrix.
(b) All the principal minors of A are nonnegative.
(c) A is nonnegative stable, *i.e.*, the real parts of the eigenvalues of A are nonnegative.
(d) There exist a permutation matrix P, a lower triangular matrix L with nonnegative diagonal entries, and an upper triangular matrix U with nonnegative diagonal entries, such that $P^T AP = LU$.
(e) There exists a positive diagonal matrix $D = \mathrm{diag}(d_1, \ldots, d_n)$ such that AD is diagonally dominant, *i.e.*, $d_i a_{ii} \geq \sum_{j \neq i} -d_j a_{ij}$.

1.24 Show that if an irreducible $n \times n$ matrix A is a singular M-matrix, then rank $A = n - 1$.

1.25 Show that if A is a singular $n \times n$ M-matrix, then there exists a positive diagonal matrix D such that $B = AD$ satisfies

$$b_{ii} = \sum_{\substack{j=1 \\ j \neq i}}^{n} -b_{ij}, \quad i = 1, \ldots, n.$$

Notes. There are several texts on nonnegative matrices and their applications. For example [Bapat and Raghavan (1997)], [Berman and Plemmons

(1994)], [Gantmacher (1959)], [Horn and Johnson (1985)], [Minc (1988)], [Seneta (1973)] and [Varga (2000)].

Corollary 1.5 was proved by Perron in 1907 and Theorems 1.12 and 1.13 were proved by Frobenius in 1912. Together they form the classical Perron-Frobenius Theorem.

References on totally nonnegative matrices include [Ando (1987)], [Gantmacher and Krein (2002)], [Gasca and Michelli (1996)] and [Karlin (1968)]. Theorem 1.14 is due to [Cryer (1976)]. In [Gantmacher and Krein (2002)] totally nonnegative matrices are called completely positive (see the notes of Section 2.1).

The M in M-matrices stands for Minkowsky. A symmetric nonsingular M-matrix is called a *Stieltjes matrix*. In other words, a Stieltjes matrix is a positive definite matrix in which the off-diagonal entries are nonpositive. The proof of Theorem 1.16 is taken from [Drew, Johnson and Loewy (1994)]. More on M-matrices can be found, *e.g.* in [Berman and Plemmons (1994)] and in [Horn and Johnson (1991)].

1.4 Schur complements

Definition 1.15 Let $A[\alpha]$ be a nonsingular principal submatrix of A. Then the *Schur complement* of $A[\alpha]$ in A, $A/A[\alpha]$, is the matrix

$$A(\alpha) - A(\alpha](A[\alpha])^{-1}A[\alpha].$$

For example, if A has block form

$$A = \begin{pmatrix} B & C \\ D & E \end{pmatrix}, \tag{1.7}$$

where E is nonsingular, then $A/E = B - CE^{-1}D$. If $A[\alpha]$ is not the right lower corner of A, we can use simultaneous permutation of the rows and columns of A to transfer $A[\alpha]$ to that corner. We therefore assume from now on that A is an $n \times n$ matrix of the form (1.7), and $A[\alpha] = E$.

Example 1.8 Let

$$A = \begin{pmatrix} 1 & 2 & 3 & 4 \\ 5 & 6 & 7 & 8 \\ 9 & 10 & 11 & 12 \\ 13 & 14 & 15 & 16 \end{pmatrix},$$

$$E_1 = \begin{pmatrix} 11 & 12 \\ 15 & 16 \end{pmatrix},$$

and $E_2 = (16)$. Then

$$\begin{aligned} A/E_1 &= \begin{pmatrix} 1 & 2 \\ 5 & 6 \end{pmatrix} - \begin{pmatrix} 3 & 4 \\ 7 & 8 \end{pmatrix} \begin{pmatrix} 11 & 12 \\ 15 & 16 \end{pmatrix}^{-1} \begin{pmatrix} 9 & 10 \\ 13 & 14 \end{pmatrix} \\ &= \begin{pmatrix} 0 & 0 \\ 0 & 0 \end{pmatrix}, \end{aligned}$$

and

$$\begin{aligned} A/E_2 &= \begin{pmatrix} 1 & 2 & 3 \\ 5 & 6 & 7 \\ 9 & 10 & 11 \end{pmatrix} - \begin{pmatrix} 4 \\ 8 \\ 12 \end{pmatrix} \left(\frac{1}{16}\right) (13 \quad 14 \quad 15) \\ &= -\frac{1}{4} \begin{pmatrix} 9 & 6 & 3 \\ 6 & 4 & 2 \\ 3 & 2 & 1 \end{pmatrix}. \end{aligned}$$

In the case that $a_{ii} \neq 0$ we also denote $A/A[i]$ by $A/[i]$.

Theorem 1.17 *Let A be as in* (1.7). *Then* $\det A = \det(A/E) \det E$ *and* $\operatorname{rank} A = \operatorname{rank}(A/E) + \operatorname{rank} E$.

Proof. Consider the equality

$$\begin{pmatrix} I & -CE^{-1} \\ 0 & I \end{pmatrix} \begin{pmatrix} B & C \\ D & E \end{pmatrix} = \begin{pmatrix} B - CE^{-1}D & 0 \\ D & E \end{pmatrix}. \tag{1.8}$$

The matrix on the right hand side of (1.8) is

$$\begin{pmatrix} A/E & 0 \\ D & E \end{pmatrix},$$

and by (1.8) it is row equivalent to A. Thus

$$\text{rank} \begin{pmatrix} A/E & 0 \\ D & E \end{pmatrix} = \text{rank}\,(A/E) + \text{rank}\,E.$$

Since

$$\det \begin{pmatrix} I & -CE^{-1} \\ 0 & I \end{pmatrix} = 1,$$

(1.8) implies that

$$\det A = \det \begin{pmatrix} A/E & 0 \\ D & E \end{pmatrix} = \det(A/E)\det E.$$

$\square$

The next proposition provides a formula for the elements of A/E.

Proposition 1.13 *Let A be an $n \times n$ matrix, and let $E = A[\alpha]$ be a nonsingular principal submatrix of A. Then for $i, j \in \{1, \ldots, n\} \setminus \alpha$,*

$$(A/E)_{ij} = \frac{\det A[\{i\} \cup \alpha | \{j\} \cup \alpha]}{\det E}.$$

Proof. Observe that $(A/E)_{ij} = A[\{i\} \cup \alpha | \{j\} \cup \alpha]/E$, and use Theorem 1.17. $\square$

Schur complements have a nice quotient property:

Theorem 1.18 *If*

$$M = \begin{pmatrix} A & B \\ C & D \end{pmatrix} \quad and \quad D = \begin{pmatrix} E & F \\ G & H \end{pmatrix},$$

and if D and H are nonsingular, then D/H is nonsingular principal submatrix of M/H, and $M/D = (M/H)/(D/H)$.

Proof. Suppose $D = M[\alpha]$, where $\alpha = \{n-k+1, \ldots, n\}$, and $H = D[\beta]$, where $\beta = \{n-l+1, \ldots, n\} \subseteq \alpha$. By Proposition 1.13, D/H is a principal submatrix of M/H; $D/H = (M/H)[\alpha \setminus \beta]$. By Theorem 1.17, $\det(D/H) = \det D/\det H$, hence D/H is nonsingular.

Let $U = (\det H)(M/H)$. The elements of U are bordered minors of H in M. Hence by Sylvester's Determinantal Identity (1.2), for every $\gamma, \delta \subseteq \{1, \ldots, n-l\}$ such that $|\gamma| = |\delta| = p$,

$$\det U[\gamma|\delta] = (\det H)^{p-1} \det M[\gamma \cup \beta | \delta \cup \beta]. \tag{1.9}$$

Now, for every $i, j \in \{1, \ldots, n - k\}$,

$$
\begin{aligned}
((M/H)/(D/H))_{ij} &= \frac{\det(M/H)[\{i\} \cup (\alpha \setminus \beta)|\{j\} \cup (\alpha \setminus \beta)]}{\det(D/H)} \\
&= \frac{(\det H)\det(M/H)[\{i\} \cup (\alpha \setminus \beta)|\{j\} \cup (\alpha \setminus \beta)]}{\det D} \\
&= \frac{\det U[\{i\} \cup (\alpha \setminus \beta)|\{j\} \cup (\alpha \setminus \beta)]}{(\det H)^{k-l}\det D}
\end{aligned}
$$

But by (1.9),

$$
\det U[\{i\} \cup (\alpha \setminus \beta)|\{j\} \cup (\alpha \setminus \beta)] = (\det H)^{k-l}\det M[\{i\} \cup \alpha|\{j\} \cup \alpha],
$$

Hence

$$
((M/H)/(D/H))_{ij} = \frac{\det M[\{i\} \cup \alpha|\{j\} \cup \alpha]}{\det D} = (M/D)_{ij}. \qquad \square
$$

We are particularly interested in the case that A is a real positive semi-definite matrix. In that case, we can extend the notion of Schur complement to the case that $A[\alpha]$ is singular, using the Moore-Penrose generalized inverse.

Definition 1.16 Let $A[\alpha]$ be a principal submatrix of a real positive semidefinite matrix A. Then the *generalized Schur complement* of $A[\alpha]$ in A is the matrix

$$
A(\alpha) - A(\alpha](A[\alpha])^{\dagger}A[\alpha].
$$

It is also denoted by $A/A[\alpha]$.

Observe that no confusion should occur, since $(A[\alpha])^{\dagger} = (A[\alpha])^{-1}$ when $A[\alpha]$ is nonsingular. As before, it would be convenient to assume that $A[\alpha]$ is the lower right corner of A. Hence we assume that A is in block form

$$
A = \begin{pmatrix} B & C \\ C^T & E \end{pmatrix}. \tag{1.10}
$$

Since A is positive semidefinite, cs $\begin{pmatrix} C^T & E \end{pmatrix} = $ cs E. Hence cs $C^T \subseteq$ cs E, and by Exercise 1.6(e) this implies that

$$
EE^{\dagger}C^T = C^T. \tag{1.11}
$$

This equality plays an essential role in proving the properties of the generalized Schur complement.

Theorem 1.19   *If A is a real positive semidefinite matrix in block form (1.10), then the matrices $A/E = B - CE^\dagger C^T$ and*

$$\begin{pmatrix} B - A/E & C \\ C^T & E \end{pmatrix}$$

are positive semidefinite.

Proof.   Using (1.11), it is easy to check that the following equality holds:

$$\begin{pmatrix} I & -CE^\dagger \\ 0 & I \end{pmatrix} \begin{pmatrix} B & C \\ C^T & E \end{pmatrix} \begin{pmatrix} I & 0 \\ -E^\dagger C^T & I \end{pmatrix} = \begin{pmatrix} B - CE^\dagger C^T & 0 \\ 0 & E \end{pmatrix}$$

Hence, by Propositions 1.2 and 1.3, A/E is positive semidefinite.

Consider now the second matrix. Since E is positive semidefinite, it has a square root and so does $E^\dagger$. Use the fact that $\operatorname{cs} C^T \subseteq \operatorname{cs} E = \operatorname{cs} \sqrt{E}$ and therefore $\sqrt{E}\sqrt{E^\dagger}C^T = \sqrt{E}\sqrt{E}^\dagger C^T = C^T$ to get that

$$\begin{pmatrix} B - A/E & C \\ C^T & E \end{pmatrix} = \begin{pmatrix} CE^\dagger C^T & C \\ C^T & E \end{pmatrix}$$

$$= \begin{pmatrix} C\sqrt{E^\dagger} \\ \sqrt{E} \end{pmatrix} \begin{pmatrix} \sqrt{E^\dagger}C^T & \sqrt{E} \end{pmatrix}.$$

By Theorem 1.10, this matrix is also positive semidefinite.   □

An equivalent to Theorem 1.10 also holds for generalized Schur complements of a positive semidefinite matrix.

Theorem 1.20   *Let A be a positive semidefinite matrix in block form (1.10). Then*

$$\det A = \det(A/E) \det E \quad \text{and} \quad \operatorname{rank} A = \operatorname{rank}(A/E) + \operatorname{rank} E,$$

and

$$\operatorname{rank} \begin{pmatrix} B - A/E & C \\ C^T & E \end{pmatrix} = \operatorname{rank} E.$$

Proof.   Prove the first part by imitating the proof of Theorem 1.17 and using (1.11).

For the proof of the second part, use again (1.11) to show that

$$\begin{pmatrix} B - A/E & C \\ C^T & E \end{pmatrix} = \begin{pmatrix} CE^\dagger C^T & C \\ C^T & E \end{pmatrix}$$

$$= \begin{pmatrix} C \\ E \end{pmatrix} \begin{pmatrix} E^\dagger C^T & I \end{pmatrix}.$$

This implies that

$$\text{rank} \begin{pmatrix} B - A/E & C \\ C^T & E \end{pmatrix} \leq \text{rank} \begin{pmatrix} C \\ E \end{pmatrix} = \text{rank } E.$$

And since the reverse inequality holds trivially, we get the desired rank equality. $\square$

The special case when A is an $n \times n$ positive semidefinite matrix and $E = (a_{nn})$, $a_{nn} \neq 0$, deserves special attention. In this case

$$\begin{pmatrix} B - A/E & C \\ C^T & E \end{pmatrix} = \mathbf{v}\mathbf{v}^T,$$

where $\mathbf{v}$ is the last column of A multiplied by $1/\sqrt{a_{nn}}$. It is a rank 1 matrix, and $A = (A/[n] \oplus 0) + \mathbf{v}\mathbf{v}^T$. By repeating the procedure for $A/[n]$, and then over and over again, one obtains a representation of A as a sum of rank A positive semidefinite matrices of rank 1 (and a factorization $A = UU^T$, where U is upper triangular — see Exercise 1.29).

Exercises

1.26   Prove that if the matrix A is an M-matrix, and E is a nonsingular principal submatrix of A, then A/E is an M-matrix.

1.27   Prove that if the matrix A is totally nonnegative, and E is a nonsingular principal submatrix of A, then A/E is also totally nonnegative.

1.28   Show that if A is as in (1.7) and E is nonsingular, then

$$\text{rank} \begin{pmatrix} CE^{-1}D & C \\ D & E \end{pmatrix} = \text{rank } E.$$

1.29   Use Schur complements to prove by induction that if A is positive semidefinite, then $A = UU^T$ for some upper triangular matrix U, and $A = LL^T$ for some lower triangular matrix L.

Combine this proof with Exercise 1.26 to deduce that if A is a symmetric nonsingular M-matrix, then $A = LL^T$ for some lower triangular M-matrix L.

Notes.   The name "Schur complement" was coined by Emilie Haynsworth in [Haynsworth (1968)]. The results on the relations between a matrix and its Schur complement are due to [Schur (1917)]. The quotient property (Theorem 1.18) was proved in [Crabtree and Haynsworth (1969)]. An interesting discussion of properties and applications of the Schur complement is carried out in [Cottle (1971)].

1.5   Graphs

In this section we establish graph terminology and notations as used in this book, and quote some relevant results. As the list is quite long, we divide this section into subsections, starting, naturally, with the basics.

Graphs and subgraphs

Definition 1.17   A *graph* G is a pair $(V(G), E(G))$ in which $V(G)$ is a finite set, whose elements are called *vertices,* and $E(G)$ is a finite set of *edges,* where each edge is an unordered pair of distinct vertices.

We use drawings to represent graphs. Each vertex is marked by a dot, and each edge by a line connecting exactly two of these dots.

Example 1.9   If

$$V(G) = \{a, b, c, d\} \ \text{ and } \ E(G) = \{\{a, b\}, \{b, c\}, \{c, d\}, \{b, d\}\},$$

then G can be drawn like that:

(This is of course just one of many possible drawings — we can place the dots wherever we wish, and draw the lines curved.)

The graphs we are talking about are usually called *simple graphs,* but we omit the "simple", since we will not be using other types of graphs, such as graphs with multiple edges (that is, several identical pairs of vertices, to be drawn like that: ⪗) and/or loops (that is, pairs of identical vertices, each to be drawn as ⬭).

If $e = \{u, v\}$ is an edge of G we say that u and v are the *vertices* or the *ends* of e, that u and v are *adjacent* or *neighbors*, that e is *incident* with u (and with v), and that u is *incident* with e (as is v). Two edges are *adjacent* if they have a common vertex. The *degree* of a vertex $v \in V(G)$ is the number of vertices adjacent to it; it is denoted by $d(v)$. A vertex v of degree 1 is called a *pendant* vertex. A graph in which every vertex is adjacent to every other edge is a *complete graph*.

Two graphs, $G_1 = (V(G_1), E(G_1))$ and $G_2 = (V(G_2), E(G_2))$, are *isomorphic* if there is a one-to-one function φ from $V(G_1)$ onto $V(G_2)$ such that $\{\varphi(u), \varphi(v)\}$ is an edge of G_2 if and only if $\{u, v\}$ is an edge of G_1. Two isomorphic graphs can be drawn exactly the same way, up to the labelling of the vertices. Since we will be mostly interested in properties of graphs which do not depend on the these labels, isomorphic graphs will often be treated as one and the same. For example, we denote any complete graph on n vertices by K_n.

K_4

The *complement* $\overline{G}$ of a graph G has the same vertices as G and two vertices in $V(\overline{G})$ are adjacent in $\overline{G}$ if and only if they are not adjacent in G.

G $\overline{G}$

A graph H is a *subgraph* of G if $V(H)$ is a subset of $V(G)$ and $E(H)$ is a subset of $E(G)$. We denote it by $H \subseteq G$, and say also that G *contains* H. If $V \subseteq V(G)$, the graph H such that $V(H) = V$ and $E(H)$ consists of all the edges of G that have both ends in V is the subgraph of G *induced* by V. Such a subgraph of G is called an *induced subgraph*. If H is a subgraph of G such that $V(H) = V(G)$, we say that H is a *spanning subgraph* of G.

Example 1.10 In the following drawing, H_1 is an induced subgraph of G, H_2 is a spanning subgraph of G, and H_3 is a subgraph of G which is neither induced nor spanning.

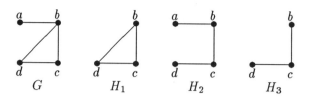

If v is a vertex of G, we denote by $G - v$ the subgraph of G induced by $V(G) \setminus \{v\}$. That is, $G - v$ is the graph obtained by deleting from G the vertex v along with all edges incident with it. If e is an edge of G, we denote by $G - e$ the graph obtained by removing e from G (that is, from $E(G)$). If u and v are vertices of G, and $e = \{u, v\}$ is not an edge of G, we denote by $G + e$ the graph obtained from G by adding the edge e.

If G_1 and G_2 are subgraphs of the same graphs G, the *union* $G_1 \cup G_2$ of the two subgraphs is the subgraph of G with vertex set $V(G_1) \cup V(G_2)$ and edge set $E(G_1) \cup E(G_2)$. G_1 and G_2 are *disjoint* if $V(G_1) \cap V(G_2) = \emptyset$. If G_1 and G_2 have at least one vertex in common, their *intersection* $G_1 \cap G_2$ has vertex set $V(G_1) \cap V(G_2)$ and edge set $E(G_1) \cap E(G_2)$.

A *clique* is a subset of $V(G)$ which induces a complete subgraph of G. It is *maximal* if it is not a proper subset of any other clique.

A sequence of edges $\{v_0, v_1\}, \{v_1, v_2\}, \ldots, \{v_{m-1}, v_m\}$ is called a *walk of length* m from v_0 to v_m. The walk is *closed* if $v_0 = v_m$ and *open* if $v_0 \neq v_m$. Such a walk from v_0 to v_m is a *path* if the vertices $v_1, \ldots, v_m$ are distinct. The vertices $v_1, \ldots, v_{m-1}$ are the *internal vertices* of this path. A *cycle* is a closed path. A *chord* of a cycle (of length > 3) is an edge which connects two nonconsecutive vertices of the cycle. A graph which is a (chordless) cycle of length n is called an *n-cycle* and denoted by C_n. The graph $C_3 = K_3$ is called a *triangle*.

C_6

If u and v are vertices in G, and there is a path in G connecting u and v, then we say that u and v are *connected* in G, and the *distance between* u *and* v *in* G, $d_G(u, v)$, is the length of the shortest path in G between u and v. A graph G is *connected* if any two distinct vertices in G are connected. Otherwise G is *disconnected*. It is easy to see that if G is disconnected, then $G = G_1 \cup \ldots \cup G_k$ for some k, where $G_1, \ldots, G_k$ are pairwise disjoint induced subgraphs of G, and each of them is connected. The subgraphs $G_1, \ldots, G_k$ are the *components* of G.

A vertex v of a graph G is a *cut vertex* of G if $G - v$ has more components than G. If G is connected, then $v \in V(G)$ is a cut vertex of G if and only if $G = G_1 \cup G_2$ where both G_1 and G_2 are connected, and the two subgraphs have only the vertex v in common. A *block* is a connected graph with no cut vertices. A *block of a graph* G is a subgraph of G, which is a block, and is not properly contained in any other block subgraph of G. A graph is k-*connected* if the deletion of any $k - 1$ vertices, together with their incident edges, leaves a connected graph with at least one edge. In particular, a 2-connected graph is a block on 3 or more vertices.

Example 1.11   The graph in the next drawing is connected, u and v are cut vertices. The graph has four blocks — a triangle, two single edges, and a 4-cycle.

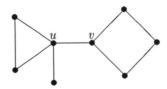

We will need the following theorem:

Theorem 1.21   *If G is 2-connected and u and v are vertices of G, then there exist at least two paths connecting u and v, which have no common internal vertex.*

Special types of graphs

We describe here some families of graphs. First — trees. A *tree* is a connected graph that contains no cycles.

Example 1.12   There are three nonisomorphic trees on 5 vertices:

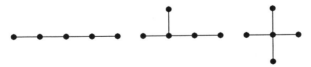

Two basic results concerning trees are:

Theorem 1.22   *A connected graph G is a tree if and only if $|E(G)| = |V(G)| - 1$.*

Theorem 1.23   *If G is a tree and $|V(G)| > 1$, then G has at least two pendant vertices.*

A graph G is *bipartite* if $V(G)$ can be partitioned into two subsets X and Y, such that each edge has one vertex in X and one vertex in Y. Equivalently, a bipartite graph is a graph that contains no cycle of odd length (no *odd cycle*). In particular, every tree and every even cycle are bipartite, and every odd cycle is not. A *complete bipartite* graph $K_{m,n}$ is a bipartite graph in which X has n vertices, Y has m vertices, and every vertex in X is adjacent to each of the vertices in Y.

$$K_{3,2}$$

While trees are bipartite graphs, bipartite graphs are part of a still larger family — the triangle free graphs. A graph G is *triangle free* if, as the name suggests, G contains no triangles. An odd cycle on 5 vertices or more is an example of a triangle free graph, which is not bipartite.

We will need the following result on triangle free graphs:

Theorem 1.24   *If G is a triangle free graph such that $|V(G)| = n$, then*

$$|E(G)| \le \lfloor n^2/4 \rfloor.$$

Equality holds if and only if $G = K_{n/2,n/2}$ (if n is even), or $G = K_{(n-1)/2,(n+1)/2}$ (if n is odd).

At the other end of the spectrum are chordal graphs: A graph G is *chordal* if each cycle in G of length greater than 3 has a chord. So G is chordal if it has no induced cycle of length greater than 3. Any complete graph is, of course, chordal, as is every tree. Another example of a chordal graph is the graph we denote by T_n, which is important in the study of completely positive matrices. The graph T_n consists of $n - 2$ triangles with a common base. That is, T_n is a graph on n vertices, where two vertices u and v are adjacent, and all other vertices are adjacent to both u and v and to no other vertex.

T_5

A special type of chordal graph is a *block-clique graph*, a connected graph in which each block is a complete graph.

Example 1.13   The drawing shows an example of a block-clique graph:

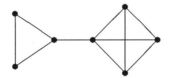

An important property of chordal graphs is the existence of a special ordering of the vertices:

Theorem 1.25   *A graph G on n vertices is chordal if and only if there is an ordering $v_1, \ldots, v_n$ of $V(G)$ such that for each $i = 1, \ldots, n-1$ the set*

$$\{v_j \mid v_j \text{ is adjacent to } v_i \text{ and } j > i\}$$

is a clique.

Such an ordering of the vertices is called a *perfect elimination ordering* (for reasons which will be explained in the notes at the end of the section).

Example 1.14   Here is a perfect elimination ordering of the vertices of T_5 (one of several possible):

For the next class of graphs we need to introduce first two important numbers associated with a graph, the chromatic number and the clique number. A graph G is *k-colorable* if $V(G)$ can be partitioned into k disjoint subsets, $V_1, \ldots, V_k$, such that adjacent vertices do not belong to the same V_i. We can think of it as coloring the vertices of G in k colors, so that any two adjacent vertices have different colors.

Example 1.15 T_5 is 3-colorable. Clearly, three colors are needed to color the vertices $3, 4, 5$, but 1 and 2 may have the same color as 3.

The smallest k such that G is k-colorable is called the *chromatic number* of G and is denoted by $\chi(G)$. For example, $\chi(T_n) = 3$ for any n. If G contains K_p as a subgraph, its chromatic number is at least p. If G has at least one edge, its chromatic number is 2 if and only if G is bipartite. If G has n vertices, then $1 \leq \chi(G) \leq n$. The upper bound is obtained by K_n and the lower bound by the graph with no edges. A better upper bound for the chromatic number is given in the following theorem.

Theorem 1.26   *If G is not K_n or an odd cycle C_{2s+1}, then $\chi(G)$ is bounded by the maximal degree of the vertices of G.*

The *clique number* of a graph G is the cardinality of the largest clique; it is denoted by $\omega(G)$. For example $\omega(T_n) = 3$ and $\omega(G) = 2$ for any triangle free graph G with at least one edge. Clearly, $\chi(G) \geq \omega(G)$.

A graph is *perfect* if $\chi(H) = \omega(H)$ for every induced subgraph H of G. Bipartite graphs are obviously perfect. Though it is not as obvious, chordal graphs also turn out to be perfect. An odd cycle is an example of a graph which is not perfect: $\chi(C_{2s+1}) = 3$ while $\omega(C_{2s+1}) = 2$. Claude Berge conjectured that a graph G is perfect if and only if neither G nor $\overline{G}$ contains an odd cycle of length greater than 3. For the state of this conjecture see the notes at the end of the section. The following weaker result is known to hold:

Theorem 1.27   *The complement of a perfect graph is perfect.*

The *line graph* of a graph G has the edges of G as its vertices, and two such vertices are adjacent if and only if they are adjacent edges of G. If H is the line graph of G, we say that G is the *root* of H.

Example 1.16   The figure shows a graph and its line graph.

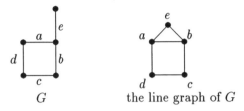

G the line graph of G

A graph G is said to be *line perfect* if its line graph is perfect. The next theorem gives a characterization of line perfect graphs.

Theorem 1.28   *A graph is line perfect if and only if it contains no odd cycle of length greater than 3.*

It turns out that these graphs have a special role in the theory of completely positive matrices (which seems to have nothing to do with their being line perfect).

Graphs and matrices

Matrices are used in graph theory to "store" the specifications of a graph, and graphs are used in matrix theory as a way to describe the zero pattern of a matrix.

We associate a matrix with every graph: Given a graph G with n vertices, label the vertices $1, 2, \ldots, n$. The *adjacency matrix*, $N(G) = (n_{ij})$, of G is the symmetric $0 - 1$ matrix where $n_{ij} = n_{ji} = 1$ if and only if i and j are adjacent.

Example 1.17   The adjacency matrix of K_n is $J_n - I_n$. The adjacency matrix of C_5, when the vertices are numbered consecutively, is

$$\begin{pmatrix} 0 & 1 & 0 & 0 & 1 \\ 1 & 0 & 1 & 0 & 0 \\ 0 & 1 & 0 & 1 & 0 \\ 0 & 0 & 1 & 0 & 1 \\ 1 & 0 & 0 & 1 & 0 \end{pmatrix}.$$

Another matrix associated with a graph is the Laplacian. The *Laplacian matrix* $L(G)$ of a graph G with vertices $1, 2, \ldots, n$ is $D - N(G)$, where

$D = \text{diag}(d_1, \ldots, d_n)$ is the diagonal matrix with $d_i = $ the degree of i. Observe that for any graph G, $L(G)$ is a singular M-matrix.

Example 1.18

$$L(C_5) = \begin{pmatrix} 2 & -1 & 0 & 0 & -1 \\ -1 & 2 & 1 & 0 & 0 \\ 0 & -1 & 2 & -1 & 0 \\ 0 & 0 & -1 & 2 & -1 \\ -1 & 0 & 0 & -1 & 2 \end{pmatrix}.$$

The classical Matrix–Tree Theorem gives the number of spanning trees of a graph. It is a nice demonstration of the relationship between graphs and matrices.

Theorem 1.29 $\text{adj}\, L(G) = kJ$ *where* k *is the number of spanning trees of* G.

In the reverse direction, we associate with each matrix a graph. For general square matrices, it is natural to use directed graphs. A *directed graph (digraph)* D consists of a finite set $V(D)$ of *vertices*, and a set $A(D)$ of ordered pairs of (not necessarily distinct) vertices, called *arcs*. It is drawn with dots for the vertices and arrows for the arcs. In a digraph the analogue of a path is a *directed path* — a sequence of arcs $(v_0, v_1), (v_1, v_2), \ldots, (v_{m-1}, v_m)$ where $v_0, \ldots, v_{m-1}$ are distinct vertices. A digraph D is *strongly connected* if for every two vertices of D there is a directed path from u to v and a directed path from v to u.

Given an $n \times n$ matrix A, the *directed graph of* A, $D(A)$, has n vertices, $1, 2, \ldots, n$, and an arc from i to j if and only if $a_{ij} \neq 0$.

Example 1.19   If

$$A = \begin{pmatrix} 1 & 2 & 0 & 0 \\ 0 & 0 & 3 & 0 \\ 0 & 4 & 0 & 5 \\ 0 & 0 & 0 & 0 \end{pmatrix},$$

then $D(A)$ is

It can be shown that a matrix A is irreducible if and only if $D(A)$ is strongly connected. More can be said if A is nonnegative:

Theorem 1.30 *Let A be an $n \times n$ irreducible nonnegative matrix. Let S_i be the set of all lengths of directed cycles through i in $D(A)$, and let h_i be the greatest common divisor of all the numbers in S_i. Then $h_1 = h_2 = \ldots = h_n = h$, and h is the index of cyclicity of A, i.e., the number of eigenvalues of A of modulus $\rho(A)$.*

The main application of graph theory in this book will be to symmetric matrices. In this case we use graphs instead of digraphs. The graph of an $n \times n$ symmetric matrix A is denoted $G(A)$. The vertices of $G(A)$ are $1, 2, \ldots, n$, and i and j are adjacent in $G(A)$ if and only if $i \neq j$ and $a_{ij} \neq 0$.

Example 1.20 If

$$A = \begin{pmatrix} 1 & 2 & 0 & 0 \\ 2 & 0 & 3 & 4 \\ 0 & 3 & 5 & 6 \\ 0 & 4 & 6 & 0 \end{pmatrix},$$

then $G(A)$ is

If A is a symmetric matrix with $G(A) = G$, we say that A is a *symmetric matrix realization* of G. If A is also nonnegative then it is a *symmetric nonnegative realization* of G. Note that in the graph of A there is no indication which diagonal entries of A are zero, and which are not. It is easy to see that a symmetric matrix is irreducible if and only if $G(A)$ is connected. If P is a permutation matrix, $G(PAP^T)$ is isomorphic to $G(A)$, since simultaneous permutation of rows and columns changes only the labels of the vertices of the graph.

Exercises

1.30   Let G be a 2-connected graph, V a proper subset of $V(G)$, $|V| \geq 2$, and v a vertex of G which is not in V. Show that there exist distinct vertices u and w in V and a path from u through v to w all of whose internal vertices are not in V.

1.31   Show that if G is a connected graph which is not a block, then G has a block which contains only one cut vertex (actually, at least two such blocks).

1.32   Prove Theorem 1.24.
Hint: Prove separately for even n and for odd n, by induction.

1.33   Show that if G is chordal, and for some $e \in E(G)$ the graph $G - e$ is also chordal, then there exits a unique maximal clique K in G such that both ends of e lie in K.

1.34   Show that if $G \neq K_n$ is a chordal graph on n vertices, then there exists an edge $e \notin E(G)$ such that $G + e$ is also chordal.
Hint: In the case that G is connected, let $v_1, \ldots, v_n$ be a perfect elimination ordering of $V(G)$. Let v_i be the first vertex adjacent to v_1 which has an adjacent vertex v_j that is not adjacent to v_1. Let $e = \{v_1, v_j\}$.

1.35   Use Theorem 1.29 to prove Cayley's Formula: The number of spanning trees of K_n is n^{n-2}.

1.36   Show that if every subgraph of G is perfect, then G contains no odd cycle of length greater than 3.

Notes.   Our favorite basic graph theory book is [Bondy and Murty (1976)]. A useful reference on perfect graphs and in particular chordal graphs and their applications is [Golumbic (1980)]. Theorem 1.21 is a special case of Menger's Theorem, which says that if a graph G with at least $k+1$ vertices is k-connected, then every two distinct vertices are connected by at least k paths which share no internal vertex. Theorem 1.24 was proved by Mantel in 1907. It is a special case of Turán's Theorem which gives a sharp upper bound for the number of edges in a graph which does not contain a K_p.

Theorem 1.25 was proved by Rose in 1970. Perfect elimination ordering of $G(A)$ is used in reordering the rows and columns of a matrix A in such a way that no nonzero elements are created during Gauss elimination (see also [Golumbic (1980)]). The term block-clique graphs is due to [Johnson and Smith (1996)]. Other names for these graphs include Husimi trees (see [Johnson and Smith (1996)]) and 1-chordal graphs (see [Drew, Johnson, Kilner and McKay (2000)]).

Theorem 1.26 was proved by Brooks in 1941 and Theorem 1.27 was proved by Lovász in 1972. Before that it was known as the Weak Perfect Graph Conjecture of Berge. The Strong Perfect Graph Conjecture was also conjectured by Berge in 1961. This is the conjecture that says that a graph G is perfect iff neither G nor its complement has an induced subgraph which is a cycle of odd length greater than 3 without any chord. In May 2002, Maria Chudnovsky and Paul Seymour announced that they, building on earlier joint work with Neil Robertson and Robin Thomas, had completed the proof of the Strong Perfect Graph Conjecture. Theorem 1.28 is due to Trotter. The graphs that contain no odd cycle of length greater than 3 play a major role in the theory of complete positivity (See Section 2.5). The Strong Perfect Graph Conjecture implies that they are all the graphs with the property that all their subgraphs are perfect (one direction does not depend on the conjecture, see Exercise 1.36).

The Matrix–Tree Theorem (Theorem 1.29) was proved by Kirchhoff in 1847. Theorem 1.30 is due to Romanovsky.

1.6 Convex cones

This section contains basic results on convex cones in a finite dimensional Euclidean space V, with inner product $\langle \ , \ \rangle$. The section is divided into several subsections.

Convexity

Definition 1.18 A set K in a Euclidean space V is *convex* if for every $\mathbf{x}, \mathbf{y} \in K$ and $0 \le a \le 1$, $a\mathbf{x} + (1 - a)\mathbf{y} \in K$.

Obviously, every linear subspace of V is a convex set. Another simple example is the closed unit ball in $\mathbf{R}^n$, $B = \{\mathbf{x} \in \mathbf{R}^n \,|\, \|\mathbf{x}\| \le 1\}$. By the triangle inequality, B is convex.

Definition 1.19 A set K in a vector space V is a *cone* if for every $\mathbf{x} \in K$ and $a \ge 0$, $a\mathbf{x} \in K$.

Combining the two definitions we get:

Definition 1.20 A set K in a Euclidean space V is a *convex cone* if for every $\mathbf{x}, \mathbf{y} \in K$ and nonnegative scalars a, b, $a\mathbf{x} + b\mathbf{y} \in K$.

Example 1.21 Let $K_1 = \mathbf{R}_+^2 = \{(x, y) \in \mathbf{R}^2 \mid x \geq 0, y \geq 0\}$ and $K_2 = \{(x, y) \in \mathbf{R}^2 \mid 0 \leq y \leq x\}$.

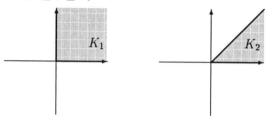

K_1 and K_2 are convex cones in $\mathbf{R}^2$. Both cones are also closed. K_1 is a particular case of a general example: For every positive integer n,

$$\mathbf{R}_+^n = \{\mathbf{x} \in \mathbf{R}^n \mid x_i \geq 0 \text{ for every } i = 1, \dots, n\}$$

is a closed convex cone.

Example 1.22 Given a unit vector $\mathbf{u} \in \mathbf{R}^n$, and an angle $0 \leq \theta < \pi/2$, let $K_{\mathbf{u},\theta}$ be the set of all vectors $\mathbf{x}$ in $\mathbf{R}^n$ such that the angle between $\mathbf{x}$ and $\mathbf{u}$ is at most θ. That is,

$$K_{\mathbf{u},\theta} = \{\mathbf{x} \in \mathbf{R}^n \mid \langle \mathbf{x}, \mathbf{u} \rangle \geq \|\mathbf{x}\| \cos \theta\}.$$

$K_{\mathbf{u},\theta}$ is a closed convex cone: It is clearly closed, and closed under multiplication by a nonnegative scalar, and if $0 \neq \mathbf{x}, \mathbf{y} \in K_{\mathbf{u},\theta}$ and $\mathbf{z} = \mathbf{x} + \mathbf{y}$, then

$$
\begin{aligned}
\langle \mathbf{z}, \mathbf{u} \rangle &= \langle \mathbf{x}, \mathbf{u} \rangle + \langle \mathbf{y}, \mathbf{u} \rangle \\
&\geq \|\mathbf{x}\| \cos \theta + \|\mathbf{y}\| \cos \theta \\
&= (\|\mathbf{x}\| + \|\mathbf{y}\|) \cos \theta \\
&\geq \|\mathbf{x} + \mathbf{y}\| \cos \theta \qquad\qquad (1.12) \\
&= \|\mathbf{z}\| \cos \theta.
\end{aligned}
$$

Hence $\mathbf{z} \in K_{\mathbf{u},\theta}$.

We will refer to cones of this type by the name *circular cones*.

For any set $S \subseteq V$, denote by $\operatorname{conv} S$ the minimal convex set containing S, and by $\operatorname{cone} S$ the minimal convex cone containing S. It is easy to see that $\operatorname{conv} S$ consists of all the convex combinations of elements in S, that is, vectors which can be represented as $\sum_{i=1}^{m} \lambda_i \mathbf{x}_i$ for some vectors $\mathbf{x}_i \in S$ and nonnegative scalars λ_i satisfying $\sum_{i=1}^{m} \lambda_i = 1$. Similarly, cone S consists of all nonnegative combinations of elements in S, that is, vectors which can be represented as $\sum_{i=1}^{m} \lambda_i \mathbf{x}_i$, where $\mathbf{x}_i \in S$ and λ_i is a nonnegative scalar, $i = 1, \ldots, m$.

The following proposition contains basic observations which follow easily from Definition 1.20:

Proposition 1.14 *Let K, K_1 and K_2 be convex cones in a Euclidean space V. Then*

(a) cl K, *the closure of K, is a convex cone.*

(b) int K, *the interior of K, is a convex cone.*

(c) *The intersection $K_1 \cap K_2$ is also a convex cone.*

(d) *The sum $K_1 + K_2 = \{\mathbf{x}_1 + \mathbf{x}_2 \mid \mathbf{x}_1 \in K_1, \mathbf{x}_2 \in K_2\}$ is a convex cone.*

Note, however, that the sum of closed convex cones is not necessarily closed:

Example 1.23 Let K_1 be the cone of vectors in $\mathbf{R}^3$ such that the angle between them and the vector $(1, 0, 1)$ is at most $\pi/4$. That is,

$$K_1 = \{(x_1, x_2, x_3) \mid x_1 \geq 0, x_3 \geq 0, 2x_1 x_3 \geq x_2^2\}.$$

Let K_2 be the x-axis in $\mathbf{R}^3$. Then

$$(0, 1, 0) = \lim_{x \to 0+} [(1/(2x), 1, x) + (-1/(2x), 0, 0)].$$

Hence $(0, 1, 0) \in \mathrm{cl}\,(K_1 + K_2)$ but $(0, 1, 0) \notin K_1 + K_2$.

Remark 1.5 If K is a convex cone in V, then $K - K$ is the minimal subspace of V containing K, and $K \cap (-K)$ is the maximal subspace of V contained in K.

Proposition 1.15 *Every convex cone K in V has a nonempty interior in $K - K$.*

Proof. Let $\mathbf{x}_1, \ldots, \mathbf{x}_k, \mathbf{y}_1, \ldots, \mathbf{y}_k$ be vectors in K such that

$$\{\mathbf{x}_1 - \mathbf{y}_1, \ldots, \mathbf{x}_k - \mathbf{y}_k\}$$

is a basis for $K - K$. Then the set $\{\mathbf{x}_1, \ldots, \mathbf{x}_k, \mathbf{y}_1, \ldots, \mathbf{y}_k\}$ spans $K - K$, and therefore contains a basis of $K - K$. So let $\{\mathbf{z}_1, \ldots, \mathbf{z}_k\} \subseteq K$ be a basis of $K - K$. Then $\{\mathbf{z} = \sum_{i=1}^{k} a_i \mathbf{z}_i \mid a_i > 0 \text{ for every } i\}$ is a nonempty open set contained in K. □

Observe that the proof relied on the fact that V (and therefore $K - K$) is finite dimensional.

Definition 1.21 The interior of a convex cone K in $K - K$ is called the *relative interior* of K, and denoted by $\operatorname{ri} K$. The set $K \setminus \operatorname{ri} K$ is the *relative boundary* of K, and it is denoted by $\operatorname{rb} K$.

By Theorem 1.14(b) $\operatorname{ri} K$ is a convex cone in $K - K$, and therefore in V.

Hyperplanes and separation

Definition 1.22 A *hyperplane* H in a Euclidean space V is a set of the form

$$H = \{\mathbf{x} \in V \mid \langle \mathbf{x}, \mathbf{u} \rangle = a\}, \tag{1.13}$$

where $\mathbf{u}$ is a nonzero vector in V, and a is a real number.

H is a hyperplane through the origin iff $a = 0$. In that case, H is a subspace of V of dimension $\dim V - 1$. If H is as in (1.13) and $a \neq 0$, then H is a translation of such a subspace; that is, $H = H_0 + \mathbf{x}_0$, where $H_0 = \{\mathbf{x} \in V \mid \langle \mathbf{x}, \mathbf{u} \rangle = 0\}$, and $\mathbf{x}_0$ is any vector in H.

Every hyperplane H defines halfspaces of V — the hyperplane in (1.13) borders two *closed halfspaces*:

$$\{\mathbf{x} \in V \mid \langle \mathbf{x}, \mathbf{u} \rangle \geq a\} \text{ and } \{\mathbf{x} \in V \mid \langle \mathbf{x}, \mathbf{u} \rangle \leq a\},$$

and two *open halfspaces*

$$\{\mathbf{x} \in V \mid \langle \mathbf{x}, \mathbf{u} \rangle > a\} \text{ and } \{\mathbf{x} \in V \mid \langle \mathbf{x}, \mathbf{u} \rangle < a\}.$$

In the case that $a = 0$, any of these halfspaces is a convex cone.

A hyperplane H is said to separate two sets S_1 and S_2 if S_1 lies in one of the halfspaces bordered by H, and S_2 lies in the opposite one. There are actually several separation variations — the halfspaces may be closed or open, the sets may be contained in H or not. We need the following:

Definition 1.23 A hyperplane H *separates properly* sets S_1 and S_2, if S_1 lies in one of the closed halfspaces bordered by H, S_2 lies in the opposite closed halfspace, and at least one of the two sets is not contained in H.

We quote here without proof two classic separation results for convex cones.

Theorem 1.31 *Let K be a closed convex cone in V, and let $\mathbf{b} \in V$, $\mathbf{b} \notin K$. Then there exists a nonzero vector $\mathbf{u} \in V$ such that $\langle \mathbf{x}, \mathbf{u} \rangle \geq 0$ for every $\mathbf{x} \in K$, and $\langle \mathbf{b}, \mathbf{u} \rangle < 0$.*

Theorem 1.32 *Let K_1 and K_2 be convex cones such that $\mathrm{ri}\, K_1 \cap \mathrm{ri}\, K_2 = \emptyset$, then there exists a hyperplane H passing through the origin, which separates K_1 and K_2 properly.*

Extreme rays

If $\mathbf{x} \in V$, then $\mathrm{cone}\,\{\mathbf{x}\} = \{a\mathbf{x} \,|\, 0 \leq a \in \mathbf{R}\}$ is the *ray generated by* $\mathbf{x}$. If $\mathbf{x} \in K$, where K is a convex cone, then the ray generated by $\mathbf{x}$ is contained in K.

Definition 1.24 Let K be a convex cone and $\mathbf{x} \in K$. If $\mathbf{x} = \mathbf{y} + \mathbf{z}$, where $\mathbf{y}, \mathbf{z} \in K$, implies that $\mathbf{y}$ and $\mathbf{z}$ are both nonnegative scalar multiples of $\mathbf{x}$, then $\mathbf{x}$ is an *extreme vector* of K, and the ray generated by $\mathbf{x}$ is an *extreme ray* of K.

Example 1.24 The extreme rays of the cone K_1 of Example 1.21 are the nonnegative x-axis, generated by $(1, 0)$, and the nonnegative y-axis, generated by $(0, 1)$. The extreme rays of the cone K_2 of Example 1.21 are the nonnegative x-axis, and the ray generated by $(1, 1)$.

Example 1.25 The extreme vectors of a circular cone $K_{\mathbf{u}, \theta}$ are the vectors which form an angle θ with $\mathbf{u}$. To see that these are indeed extreme vectors, let $\mathbf{z}$ satisfy $\langle \mathbf{z}, \mathbf{u} \rangle = \|\mathbf{z}\| \cos \theta$ and suppose $\mathbf{z} = \mathbf{x} + \mathbf{y}$, where $\mathbf{x}, \mathbf{y} \in K_{\mathbf{u}, \theta}$. Then equality necessarily holds in (1.12), and therefore one of the vectors $\mathbf{x}, \mathbf{y}$ is a nonnegative multiple of the other, and so is $\mathbf{z}$. We leave it for the reader to show that these are all the extreme vectors.

The following proposition follows easily from Definition 1.24.

Proposition 1.16 *If $K = \mathrm{cone}\, S$, then there exists a subset T of S such that the vectors in T generate all the extreme rays of K.*

The next theorem is a kind of inverse of Proposition 1.16.

Theorem 1.33 [Krein-Milman Theorem] *If T is a set of extreme vectors of a closed convex cone K which generate all the extreme rays of K, then $K = \mathrm{cl}\,(\mathrm{cone}\, T)$.*

The next theorem and its proof will be very useful in Chapter 3.

Theorem 1.34 [Carathéodory's Theorem] *If $K = \text{cone } S$ is a convex cone in a Euclidean space of dimension n, then every $\mathbf{x} \in K$ can be represented as a nonnegative combination of at most n elements of S.*

Proof. Let $\mathbf{x} \in K$. Then $\mathbf{x} = \sum_{i=1}^{m} \lambda_i \mathbf{x}_i$, for some $\mathbf{x}_1, \ldots, \mathbf{x}_m \in S$ and positive $\lambda_1, \ldots, \lambda_m$. We show that if $m > n$, then $\mathbf{x}$ may also be represented as a positive combination of less then m elements. If $m > n$, then $\mathbf{x}_1, \ldots, \mathbf{x}_m$ are linearly dependent, hence there exist real numbers μ_i, $i = 1, \ldots, m$, not all zero, such that $\mathbf{0} = \sum_{i=1}^{m} \mu_i \mathbf{x}_i$. We may assume that at least one μ_i is positive. Let

$$a = \min \left\{ \frac{\lambda_i}{\mu_i} \;\middle|\; \mu_i > 0 \right\},$$

and $\nu_i = \lambda_i - a\mu_i$, $i = 1, \ldots, m$. Then $\mathbf{x} = \sum_{i=1}^{m} \nu_i \mathbf{x}_i$, where all the ν_i are nonnegative, and at least one of them is zero. □

Remark 1.6   By the proof of Theorem 1.34, if $\mathbf{x}$ is a positive combination of m linearly dependent vectors $\mathbf{x}_1, \ldots, \mathbf{x}_m$, then there exists a linearly independent subset of $\{\mathbf{x}_1, \ldots, \mathbf{x}_m\}$ such that $\mathbf{x}$ is a positive combination of its elements.

Duality

Definition 1.25   Let S be a set in V. The set

$$S^* = \{\mathbf{y} \in V \mid \langle \mathbf{y}, \mathbf{x} \rangle \geq 0 \text{ for every } \mathbf{x} \in S\}$$

is called the *dual* of S.

Example 1.26   For K_2 of Example 1.21, $K_2^* = \{(x, y) \mid 0 \leq x, -x \leq y\}$. To see that, note that $(1, 1) \in K_2$ and $(1, 0) \in K_2$. Therefore, if $(x, y) \in K_2^*$, then $\langle (x, y), (1, 1) \rangle \geq 0$ and $\langle (x, y), (1, 0) \rangle \geq 0$, that is, $x + y \geq 0$ and $x \geq 0$. This proves that $K_2^* \subseteq \{(x, y) \mid 0 \leq x, -x \leq y\}$. To prove the reverse inclusion, suppose $(x_1, y_1) \in K_2$ and $(x_2, y_2) \in \{(x, y) \mid 0 \leq x, -x \leq y\}$. Then $\langle (x_1, y_1), (x_2, y_2) \rangle = x_1 x_2 + y_1 y_2 \geq x_1 x_2 - y_1 x_2 = (x_1 - y_1) x_2 \geq 0$.

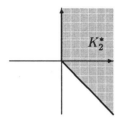

Similarly, it is easy to see that the cone K_1 of the same example is self dual, i.e., $K_1^* = K_1$.

We leave the next two examples as exercises.

Example 1.27 The dual of the circular cone $K_{\mathbf{u},\theta}$ is the circular cone $K_{\mathbf{u},\pi/2-\theta}$.

Example 1.28 If L is a subspace of V, then $L^* = L^\perp$.

The following theorem lists some basic (and easy to prove) duality properties.

Theorem 1.35 *Let S, S_1, S_2 be nonempty subsets of V. Then*

(a) S^* *is a closed convex cone.*
(b) *If $S_1 \subseteq S_2$, then $S_2^* \subseteq S_1^*$.*
(c) $S \subseteq S^{**}$.
(d) $S^* = (\operatorname{cl} S)^* = (\operatorname{cone} S)^* = (\operatorname{conv} S)^*$.
(e) $S_1^* \cap S_2^* \subseteq (S_1 + S_2)^*$, *and if $\mathbf{0} \in S_1 + S_2$, $(S_1 + S_2)^* \subseteq S_1^* \cap S_2^*$.*

Theorem 1.36 *S is a closed convex cone if and only if $S = S^{**}$.*

Proof. "If": Follows from Theorem 1.35(a).

"Only if": By Theorem 1.35(c), it is enough to show that $S^{**} \subseteq S$. Suppose $\mathbf{b} \in S^{**}$, $\mathbf{b} \notin S$. Then by Theorem 1.31 there is a vector $\mathbf{u} \in V$ such that $\langle \mathbf{u}, \mathbf{b} \rangle < 0$, while $\langle \mathbf{u}, \mathbf{x} \rangle \geq 0$ for every $\mathbf{x} \in S$. But then $\mathbf{u} \in S^*$ and thus $\mathbf{b} \notin S^{**}$, a contradiction. $\square$

Corollary 1.7 *If K is a convex cone, then $\operatorname{cl} K = K^{**}$.*

Proof. By Theorem 1.35(d) $K^* = (\operatorname{cl} K)^*$, and therefore $K^{**} = (\operatorname{cl} K)^{**}$. By Theorem 1.36, $(\operatorname{cl} K)^{**} = \operatorname{cl} K$. $\square$

When S_1 and S_2 in Theorem 1.35 are closed convex cones, statement (e) can be strengthened.

Corollary 1.8 *If K_1 and K_2 are closed convex cones then*

$$\operatorname{cl}(K_1^* + K_2^*) = (K_1 \cap K_2)^*.$$

Proof. $(K_1 \cap K_2)^* = (K_1^{**} \cap K_2^{**})^* = (K_1^* + K_2^*)^{**} = \operatorname{cl}(K_1^* + K_2^*)$. The middle equality follows from Theorem 1.35(e) and the right equality follows from Corollary 1.7, since the sum of convex cones is a convex cone. $\square$

Types of convex cones

Definition 1.26 A convex cone K in V is *pointed* if $K \cap (-K) = \{0\}$, *reproducing* if $K - K = V$, *solid* if int $K \neq \emptyset$, and *full* if it is closed, convex, pointed and solid.

Recall that $K - K$ is the minimal subspace of V containing K, and that $K \cap (-K)$ is the maximal subspace of V contained in K. Hence a pointed cone contains no subspace but $\{0\}$, and no proper subspace of V contains a reproducing cone K.

Remark 1.7 A pointed closed convex cone in V induces a partial order in V, via $\mathbf{x} \leq \mathbf{y}$ iff $\mathbf{y} - \mathbf{x} \in K$.

Proposition 1.17 *A closed convex cone in V is solid if and only if it is reproducing.*

Proof. "If": See Proposition 1.15

"Only if": Let $\mathbf{v}$ be an interior point of K. Then for every $\mathbf{x} \in V$ there is a sufficiently small positive d such that $\mathbf{u} = \mathbf{v} + d\mathbf{x} \in \operatorname{int} K$. Since $\mathbf{x} = \mathbf{u}/d - \mathbf{v}/d$, $\mathbf{x} \in \operatorname{int} K - \operatorname{int} K$. Hence $V \subseteq \operatorname{int} K - \operatorname{int} K$, K is reproducing. $\square$

The next proposition relies on Proposition 1.17, and we leave the proof as an exercise.

Proposition 1.18 *A closed convex cone K is pointed if and only if K^* is solid.*

Theorem 1.37 *If K is a closed pointed cone in V, then there is a hyperplane H separating K and $-K$, such that $H \cap K = \{0\}$.*

Proof. If K is a closed pointed cone, then there exists a pointed cone K' such that $K \setminus 0 \subseteq \operatorname{ri} K'$ (see Exercise 1.44). By Theorem 1.32 there exists a hyperplane H through the origin which separates K' and $-K'$ properly. Clearly, $H \cap \operatorname{ri} K' = \emptyset$; hence $H \cap K = \{0\}$. $\square$

Definition 1.27 A convex cone K is a *polyhedral cone* if it is finitely generated, *i.e.*, $K = \operatorname{cone} S$ for some finite set S.

$\mathbf{R}^n$ is a polyhedral cone.

Proposition 1.19 *A polyhedral cone is necessarily closed.*

Note that K is a polyhedral cone if and only if K has a finite number of extreme rays (this follows from the last proposition together with Proposition 1.16 and Theorem 1.33).

Remark 1.8 Every convex cone in $\mathbf{R}^2$ is polyhedral. But for $n \geq 3$, there are convex cones which are not polyhedral. For example, any circular cone in $\mathbf{R}^n$, $n \geq 3$, has infinitely many extreme rays, and therefore is not polyhedral.

We add the following without proof:

Theorem 1.38 *K is a polyhedral cone iff it is the intersection of finitely many closed halfspaces.*

Cones of matrices

We will be interested mostly in symmetric matrices, hence we consider $\mathcal{S}_n$ — the space of all symmetric matrices. $\mathcal{S}_n$ is a subspace of $\mathbf{R}^{n \times n}$, which is isomorphic to $\mathbf{R}^{n(n+1)/2}$. The regular inner product in $\mathbf{R}^{n \times n}$ translates in $\mathcal{S}_n$ into the following: $\langle A, B \rangle = \operatorname{trace}(AB)$. Several sets of symmetric matrices form convex cones in $\mathcal{S}_n$ (and therefore in $\mathbf{R}^{n \times n}$).

Let $\mathcal{SNN}_n$ denote the set of $n \times n$ symmetric nonnegative matrices. The following proposition is easy to prove and is left as an exercise.

Proposition 1.20 *The set $\mathcal{SNN}_n$ is a full polyhedral cone in $\mathcal{S}_n$. Its interior consists of all symmetric positive matrices. Its extreme rays are all the rays generated by E_{ii}, $i = 1, \ldots, n$, and those generated by the matrices $E_{ij} + E_{ji}$, $1 \leq i < j \leq n$. As a cone in $\mathcal{S}_n$, $\mathcal{SNN}_n$ is self dual.*

Remark 1.9 $\mathcal{SNN}_n$ is a subset of $\mathcal{NN}_n$ — the set of all nonnegative $n \times n$ matrices, which is a convex cone in $\mathbf{R}^{n \times n}$.

Let $\mathcal{PSD}_n$ denote the set of $n \times n$ positive semidefinite matrices.

Proposition 1.21 *$\mathcal{PSD}_n$ is a full cone in $\mathcal{S}_n$. The interior of $\mathcal{PSD}_n$ is the open cone $\mathcal{PD}_n$ of positive definite matrices, and its extreme vectors are the rank 1 symmetric matrices. $\mathcal{PSD}_n$ is self dual.*

Proof. $\mathcal{PSD}_n$ is a convex cone by Proposition 1.1. It is closed, since $A_k \in \mathcal{PSD}_n$ and $\lim_{k\to\infty} A_k = A$ implies that for every $\mathbf{x} \in \mathbf{R}^n$,

$$\langle A\mathbf{x}, \mathbf{x} \rangle = \lim_{k\to\infty} \langle A_k\mathbf{x}, \mathbf{x} \rangle \geq 0.$$

It is easy to see that $\mathcal{PSD}_n$ is pointed: If $\langle A\mathbf{x}, \mathbf{x} \rangle \geq 0$ and $\langle -A\mathbf{x}, \mathbf{x} \rangle \geq 0$ for every $\mathbf{x} \in \mathbf{R}^n$, then $\langle A\mathbf{x}, \mathbf{x} \rangle = 0$ for every $\mathbf{x}$, hence $A = 0$.

We show that $\text{int}\,\mathcal{PSD}_n = \mathcal{PD}_n$ (in particular, $\mathcal{PSD}_n$ is solid): Suppose A is a positive definite matrix. Then all the leading minors of A are positive, and are continuous functions of the matrix entries. Hence there exists an $\varepsilon > 0$ such that a perturbation of each entry of A by at most ε will still leave all leading principal minors positive. Then $A + \varepsilon B$ is positive definite for every matrix $B \in \mathcal{S}_n$ such that $\langle B, B \rangle \leq 1$.

Next, every rank 1 symmetric matrix is an extreme vector of $\mathcal{PSD}_n$: If $\mathbf{x}\mathbf{x}^T = A + B$, where A and B are positive semidefinite, then by Exercise 1.8, $\text{ns}\,A \cap \text{ns}\,B = \text{ns}\,\mathbf{x}\mathbf{x}^T = \{\mathbf{x}\}^\perp$. Hence either $\text{ns}\,A = \{\mathbf{x}\}^\perp$, or $\text{ns}\,A = \mathbf{R}^n$. In any case, A is a scalar multiple of $\mathbf{x}\mathbf{x}^T$. By Theorem 1.10(h), $\mathcal{PSD}_n$ is generated by the rank 1 matrices.

Finally, to see that $\mathcal{PSD}_n^* = \mathcal{PSD}_n$, note that

$$\text{trace}\,(AB) = \mathbf{e}^T A \circ B \mathbf{e}.$$

Since $A \circ B$ is positive semidefinite if A and B are (Proposition 1.7), this equality implies that $\mathcal{PSD}_n \subseteq \mathcal{PSD}_n^*$. For the reverse inclusion, note that if $A \in \mathcal{PSD}_n^*$ then $\text{trace}\,(\mathbf{x}^T A\mathbf{x}) = \text{trace}\,(A\mathbf{x}\mathbf{x}^T) \geq 0$ for every $\mathbf{x} \in \mathbf{R}^n$. $\square$

Note that $\mathcal{PSD}_n$ is not polyhedral, since it has infinitely many extreme rays.

Next we consider $\mathcal{DNN}_n$ — the set of $n \times n$ matrices which are both nonnegative and positive semidefinite.

Proposition 1.22 $\mathcal{DNN}_n$ *is a full cone in $\mathcal{S}_n$ and $\text{int}\,\mathcal{DNN}_n$ consists of the positive definite matrices which are also entrywise positive.*

Proof. $\mathcal{DNN}_n = \mathcal{SNN}_n \cap \mathcal{PSD}_n$. This implies, by the last two propositions, that $\mathcal{DNN}_n$ is a closed pointed convex cone.

$$\text{int}\,\mathcal{DNN}_n = \text{int}\,\mathcal{SNN}_n \cap \text{int}\,\mathcal{PSD}_n \neq \emptyset.$$

For example, $J + I$ is a positive definite positive matrix. $\square$

There is no general description of the extreme rays of $\mathcal{DNN}_n$. However, there are some partial results. They are stated here without proof.

Proposition 1.23 *For $n \leq 4$, the extreme vectors of $\mathcal{DNN}_n$ are the rank 1 matrices in $\mathcal{DNN}_n$.*

For $n = 5$, $A \in \mathcal{DNN}_5$ is an extreme vector of $\mathcal{DNN}_5$ if and only if rank $A = 1$, or rank $A = 3$ and $G(A) = C_5$.

We conclude this section with the set of copositive matrices.

Definition 1.28 A matrix $A \in \mathbf{R}^{n \times n}$ is *copositive* if it is symmetric and $\mathbf{x}^T A \mathbf{x}$ is nonnegative for every nonnegative $\mathbf{x} \in \mathbf{R}^n$.

Some examples of copositive matrices are the positive semidefinite matrices and the symmetric nonnegative matrices. Note that if A is a copositive matrix, then $\mathbf{e}_i^T A \mathbf{e}_i \geq 0$ implies that $a_{ii} \geq 0$ for every i.

We denote by $\mathcal{COP}_n$ the set of all $n \times n$ copositive matrices.

Proposition 1.24 *$\mathcal{COP}_n$ is a full cone.*

Proof. The proof that $\mathcal{COP}_n$ is a closed convex cone is similar to the proof that $\mathcal{PSD}_n$ is a closed convex cone, and is left as an exercise. Since $\mathcal{COP}_n$ contains the solid cone $\mathcal{PSD}_n$, $\mathcal{COP}_n$ is also solid.

Finally, if $A \in \mathcal{COP}_n \cap (-\mathcal{COP}_n)$, then $\mathbf{x}^T A \mathbf{x} = 0$ for every nonnegative vector $\mathbf{x} \in \mathbf{R}^n$. In particular, taking $\mathbf{x} = \mathbf{e}_i$ implies that $a_{ii} = 0$ for every $i = 1, \ldots, n$. This, together with the fact that $(\mathbf{e}_i^T + \mathbf{e}_j^T)A(\mathbf{e}_i + \mathbf{e}_j) = 0$, yields also $a_{ij} = 0$ for every $1 \leq i \neq j \leq n$. □

The interior of $\mathcal{COP}_n$ and its extreme rays are not fully known, but we have the following partial result:

Proposition 1.25 *The following matrices generate extreme rays of $\mathcal{COP}_n$:*

(a) E_{ii}, $i = 1, \ldots, n$.
(b) $E_{ij} + E_{ji}$, $1 \leq i < j \leq n$.
(c) $\mathbf{x}\mathbf{x}^T$, where $\mathbf{x} \in \mathbf{R}^n$ has some positive entries and some negative entries.

Proof. As mentioned above, the diagonal entries in a copositive matrix are nonnegative. Now note also that if A is copositive and $a_{ii} = 0$, then all the other entries in the i-th row (column) of A are nonnegative. To see that, take $\mathbf{x} = \mathbf{e}_i + \varepsilon \mathbf{e}_j$, for $j \neq i$. Then $\mathbf{x}^T A \mathbf{x} = 2\varepsilon a_{ij} + \varepsilon^2 a_{jj}$. Hence for every $\varepsilon > 0$, $a_{ij} + \varepsilon a_{jj} \geq 0$, which implies that $a_{ij} \geq 0$.

So if $E_{ii} = A + B$, where A and B are copositive, then for every $j \neq i$,

$$a_{jj} + b_{jj} = 0, \quad , a_{jj} \geq 0, \quad b_{jj} \geq 0.$$

Hence $a_{jj} = b_{jj} = 0$. But then for every $j \neq i$ and any k,

$$a_{kj} + b_{kj} = 0, \quad , a_{kj} \geq 0, \quad b_{kj} \geq 0,$$

which implies that $a_{kj} = b_{kj} = 0$. Hence

$$A = a_{ii}E_{ii}, \quad B = b_{ii}E_{ii}.$$

This proves that E_{ii} generates an extreme ray. The proof that $E_{ij} + E_{ji}$, $1 \leq i < j \leq n$, generates an extreme ray is similar, and is left as an exercise.

We now turn to examine matrices of the form (c). A few simple observations may be helpful: If P is a permutation matrix, then A generates an extreme ray of $\mathcal{COP}_n$ if and only if $P^T A P$ does. The same holds for any positive diagonal matrix D. Hence, it suffices to show that $\mathbf{xx}^T$ generates an extreme ray of $\mathcal{COP}_n$ when

$$x_1 = \ldots = x_k = 1, \quad x_{k+1} = \ldots = x_m = -1, \quad x_{m+1} = \ldots = x_n = 0. \tag{1.14}$$

Consider first the special case that $\mathbf{x}^T = (1, -1)$. We show that in this case $\mathbf{xx}^T$ is an extreme vector of $\mathcal{COP}_2$. Suppose

$$\mathbf{xx}^T = \begin{pmatrix} 1 & -1 \\ -1 & 1 \end{pmatrix} = A + B,$$

where A and B are copositive. Then without loss of generality we may assume that

$$A = \begin{pmatrix} a & b \\ b & c \end{pmatrix},$$

where $b < 0$ and $a, c \geq 0$. Since $\mathbf{e}^T \mathbf{xx}^T \mathbf{e} = 0$ and $\mathbf{e}^T A \mathbf{e}, \mathbf{e}^T B \mathbf{e} \geq 0$, we necessarily have $\mathbf{e}^T A \mathbf{e} = 0$. That is,

$$a + 2b + c = 0. \tag{1.15}$$

Hence for every $\mathbf{y} \in \mathbf{R}^2$

$$\mathbf{y}^T A \mathbf{y} = a y_1^2 + 2 b y_1 y_2 - (a + 2b) y_2^2 = (y_1 - y_2)(a(y_1 + y_2) + 2b y_2). \tag{1.16}$$

Now, if $a + b > 0$, let $\mathbf{y}^T = (1 - \varepsilon, 1)$. By (1.16), $\mathbf{y}^T A \mathbf{y} = -\varepsilon(2a - \varepsilon a + 2b)$ is negative when $\varepsilon > 0$ is sufficiently small, which contradicts the copositivity of A. On the other hand, if $a + b < 0$, then $\mathbf{y}^T = (1 + \varepsilon, 1)$ yields a negative $\mathbf{y}^T A \mathbf{y}$ when $\varepsilon > 0$ is sufficiently small — a contradiction again. Therefore

$a + b = 0$ and, due to (1.15), $b + c = 0$. This implies that A (and hence also B) is a scalar multiple of $\mathbf{xx}^T$.

In the general case $\mathbf{x}$ is as in (1.14) and

$$\mathbf{xx}^T = \begin{pmatrix} X_{11} & X_{12} & 0 \\ X_{12}^T & X_{22} & 0 \\ 0 & 0 & 0 \end{pmatrix},$$

where $X_{11} = J_k$, $X_{22} = J_{m-k}$, X_{12} is a $k \times (m-k)$ matrix with all entries equal to -1, and each 0 stands for a zero matrix of appropriate order. If $\mathbf{xx}^T = A + B$ where A and B are copositive, then using the same arguments as in the beginning of the proof, $a_{ij} = b_{ij} = 0$ for every i and every $j > m$. For every $i < k$ and $k < j \leq m$,

$$\mathbf{xx}^T[i,j] = \begin{pmatrix} 1 & -1 \\ -1 & 1 \end{pmatrix}$$

is an extreme matrix in $\mathcal{COP}_2$, and hence

$$A[i,j] = a_{ii} \begin{pmatrix} 1 & -1 \\ -1 & 1 \end{pmatrix}.$$

That is, $a_{ii} = a_{jj} = -a_{ij}$ for every $i < k$ and $k < j \leq m$. Combining all these equalities we get that $A = a_{11}\mathbf{xx}^T$, and this implies that $B = (1 - a_{11})\mathbf{xx}^T$. □

Example 1.29 The matrix J_n does not generate an extreme ray of $\mathcal{COP}_n$, since

$$J_n = \mathbf{xx}^T + 2\sum_{j=2}^{n}(E_{1j} + E_{j1}),$$

where $\mathbf{x}$ is the vector in $\mathbf{R}^n$ whose first entry is -1 and the other $n - 1$ entries are all equal to 1.

Remark 1.10 As mentioned above,

$$\mathcal{COP}_n \supseteq \mathcal{SNN}_n \text{ and } \mathcal{COP}_n \supseteq \mathcal{PSD}_n.$$

Since $\mathcal{COP}_n$ is a convex cone, this implies that

$$\mathcal{COP}_n \supseteq \mathcal{SNN}_n + \mathcal{PSD}_n.$$

By Proposition 1.29 is that all the extreme rays of $\mathcal{SNN}_n$ and some of the extreme rays of $\mathcal{PSD}_n$ are also extreme rays of $\mathcal{COP}_n$. However, there may be other extreme rays, and $\mathcal{COP}_n$ is not necessarily equal to $\mathcal{SNN}_n + \mathcal{PSD}_n$. Equality does hold for $n \leq 4$, but not for $n \geq 5$, as the next example shows. We leave some of the details of the example as an exercise.

Example 1.30 The matrix

$$Q = \begin{pmatrix} 1 & -1 & 1 & 1 & -1 \\ -1 & 1 & -1 & 1 & 1 \\ 1 & -1 & 1 & -1 & 1 \\ 1 & 1 & -1 & 1 & -1 \\ -1 & 1 & 1 & -1 & 1 \end{pmatrix}$$

is copositive. To see that, note that

$$\mathbf{x}^T Q \mathbf{x} = (x_1 - x_2 + x_3 + x_4 - x_5)^2 + 4x_2x_4 + 4x_3(x_5 - x_4) \quad (1.17)$$
$$= (x_1 - x_2 + x_3 - x_4 + x_5)^2 + 4x_2x_5 + 4x_1(x_4 - x_5) \quad (1.18)$$

(1.17) implies that $\mathbf{x}^T Q \mathbf{x} \geq 0$ for any nonnegative vector $\mathbf{x}$ with $x_5 \geq x_4$, and (1.18) implies that $\mathbf{x}^T Q \mathbf{x} \geq 0$ for any nonnegative vector $\mathbf{x}$ with $x_5 < x_4$.

The matrix Q generates an extreme ray of $\mathcal{COP}_5$ (see Exercise 1.50). Since Q is neither nonnegative, nor positive semidefinite, and it is extreme, it cannot be represented as a sum of a positive semidefinite matrix and a symmetric nonnegative matrix. In particular, $\mathcal{COP}_5 \neq \mathcal{PSD}_5 + \mathcal{SNN}_5$.

Exercises

1.37   Prove Proposition 1.14.

1.38   Show that $K - K$ is the minimal subspace of V containing K, and $K \cap (-K)$ is the maximal subspace of V contained in K.

1.39   Show that if $\mathbf{x}$ forms an angle less than θ with $\mathbf{u}$, then $\mathbf{x}$ is not an extreme vector of $K_{\mathbf{u},\theta}$.

1.40   Prove the statement in Example 1.27.

1.41   Prove the statement in Example 1.28.

1.42   Prove Theorem 1.35.

1.43   Prove Proposition 1.18

1.44   Prove that if K is a pointed cone, then there exists a pointed cone K' such that $K \setminus \{0\} \subseteq \mathrm{ri}\, K'$.

1.45   Show that every closed convex cone in $\mathbf{R}^2$ is polyhedral.

1.46   Prove Proposition 1.19.

1.47   Prove Proposition 1.20.

1.48   Prove the first part of Proposition 1.16: $\mathcal{COP}_n$ is a closed convex cone.

1.49   Show that for $1 \leq i < j \leq n$ the matrix $E_{ij} + E_{ji}$ generates an extreme ray of $\mathcal{COP}_n$.

1.50   Using the fact that

$$\begin{pmatrix} 1 & -1 & 1 \\ -1 & 1 & -1 \\ 1 & -1 & 1 \end{pmatrix}$$

generates an extreme ray of $\mathcal{COP}_3$, show that the matrix Q of Example 1.30 generates an extreme ray of $\mathcal{COP}_5$.

Notes.   A classical reference on convexity is [Rockafellar (1970)]. Cones in matrix theory are discussed in [Berman (1973)].

Example 1.23 is a counter example to Lemma 16.2.2 of [Hall (1986)].

Theorem 1.33 is a finite dimensional version of the Krein-Milman Theorem. Theorem 1.34 is a version of Carathéodory's Theorem for convex cones.

Proposition 1.17 is a theorem of Krein and Rutman, [Krein and Rutman (1956)].

If K is a closed convex cone in $\mathbf{R}^n$, $\pi(K)$ is defined to be the set of $n \times n$ matrices A for which $AK \subseteq K$. It is a closed convex cone. Note that $\mathcal{NN}_n = \pi(\mathbf{R}_+^n)$. The Perron-Frobenius theory of nonnegative matrices has generalizations for matrices in $\pi(K)$. See [Berman and Plemmons (1994)].

The characterization of the extreme rays of $\mathcal{DNN}_n$, $n \leq 5$, was done in [Ycart (1982)] (See also [Hamilton-Jester and Li (1996)] for this and some other results on the extreme rays of $\mathcal{DNN}_n$).

The discussion of the extreme rays of $\mathcal{COP}_n$ is based on [Hall (1986)], including Example 1.30, which is due to A. Horn. The fact that $\mathcal{COP}_n = \mathcal{SNN}_n + \mathcal{PSD}_n$ for $n \leq 4$ is due to [Diananda (1962)]. A necessary and sufficient condition for copositivity was obtained by Motzkin, see [Motzkin (1952)].

1.7   The PSD completion problem

Definition 1.29   A *partial matrix* is a "matrix" where some of the entries are not specified. We denote the unspecified entries by a question mark. A *completion* of a partial matrix is a real matrix obtained by replacing the unspecified entries by real numbers.

Example 1.31

$$\begin{pmatrix} e & -1 & ? \\ \pi & ? & 1 \\ 0 & ? & ? \end{pmatrix}$$

is a partial matrix.

Definition 1.30   A *partial symmetric matrix* A is a partial matrix in which the diagonal entries are specified, and if a_{ij} is specified, then a_{ji} is also specified and equal to a_{ji}.

Example 1.32

$$\begin{pmatrix} e & \pi & 0 \\ \pi & -1 & ? \\ 0 & ? & 1 \end{pmatrix}$$

is a partial symmetric matrix.

The reason for the requirement in Definition 1.30 that the diagonal entries be specified will be explained in the notes at the end of the section (with the aid of the exercises).

Definition 1.31   The *specification graph* of an $n \times n$ partial symmetric matrix A is a graph with vertices $\{1, 2, \ldots, n\}$, in which i and j are adjacent if and only if $i \neq j$ and a_{ij} is specified. If G is the specification graph of a partial symmetric matrix A we say that A is a *G-partial* matrix.

Example 1.33   The specification graph of the matrix in Example 1.32 is

Definition 1.32 A *partial PSD matrix* is a partial symmetric matrix in which all the fully specified principal submatrices are positive semidefinite. If A is a partial PSD matrix with specification graph G, we say that A is a *G-partial PSD* matrix. A *PSD completion* of A is a completion of A which is positive semidefinite.

Note that by the definition, the diagonal entries of a partial PSD matrix are nonnegative.

Example 1.34 Let A be the $n \times n$ partial matrix with all entries specified and equal to 1, except for a couple of unspecified entries in positions $1, n$ and $n, 1$. The matrix A has a completion which is positive semidefinite: the matrix J_n. In fact, J_n is the only possible PSD completion of A: If in such a completion B the $1n$ and $n1$ entries are equal to a, then

$$B[1, 2, n] = \begin{pmatrix} 1 & 1 & a \\ 1 & 1 & 1 \\ a & 1 & 1 \end{pmatrix}.$$

Hence $\det(B[1, 2, n]) = -(1 - a)^2$, and necessarily $a = 1$.

In fact, Example 1.34 is a special case of the following proposition, whose proof is left as an exercise:

Proposition 1.26 *If A is a partial PSD matrix with only one pair of unspecified entries, then A has a PSD completion.*

Consider the following question: Given a partial PSD matrix A, how can one tell if it can be completed to a positive semidefinite matrix? Clearly, if A is not partial PSD, it cannot be completed to a positive semidefinite matrix. But even if A is partial PSD, it does not guarantee PSD completability.

Example 1.35 Let A be a $k \times k$ partial symmetric matrix, $k > 3$, in which the diagonal, first superdiagonal and first subdiagonal entries are equal to 1, the $1k$ and the $k1$ entries are equal to 0, and all other entries are unspecified:

$$A = \begin{pmatrix} 1 & 1 & ? & & \cdots & ? & 0 \\ 1 & 1 & 1 & ? & & ? & ? \\ ? & 1 & 1 & \ddots & \ddots & & \vdots \\ & ? & \ddots & \ddots & \ddots & ? & \\ \vdots & & \ddots & \ddots & 1 & 1 & ? \\ ? & ? & & ? & 1 & 1 & 1 \\ 0 & ? & \cdots & \cdots & ? & 1 & 1 \end{pmatrix}.$$

A is partial PSD, since the fully specified principal submatrices are the diagonal entries, which are positive, and the 2×2 principal submatrices, which are either J_2 or I_2. However, A cannot be completed to a positive semidefinite matrix. To see that, observe that by a repeated use of Example 1.34, the only value that can be placed instead of any of the "?" signs is 1. But then

$$\text{rank } B[1,2] = 1 \neq 2 = \text{rank } B[1,2|1,\ldots,n],$$

a contradiction the Principal Submatrix Rank Property (Proposition 1.5) .

It is natural to define:

Definition 1.33   A graph G is *PSD completable* if any G-partial PSD matrix A can be completed to a positive semidefinite matrix.

Since positive semidefiniteness is preserved by a simultaneous permutation of rows and columns, the labelling of the vertices of a graph does not affect its PSD completability. The specification graph of the matrix in Example 1.35 is C_k, so C_k, $k > 3$, is not PSD completable. By Proposition 1.26 the graph on n vertices which is one edge short of being the complete graph is an example of a PSD completable graph.

The following theorem is a full characterization of PSD completable graphs:

Theorem 1.39   *A graph G is PSD completable if and only if it is chordal.*

Proof.   We prove first that if G is not chordal, then it is not PSD completable. If G is not chordal, then it contains an induced cycle of length $k > 3$. Construct a G-partial PSD matrix A: Let the principal submatrix corresponding to the cycle be the matrix in Example 1.26, set all other diagonal entries equal to 1, and all other specified entries equal to 0. Then A is partial PSD matrix that cannot be completed to a positive semidefinite matrix.

Now suppose G is a chordal graph on n vertices. Then there is a sequence of chordal graphs $G_0 = G, G_1, \ldots, G_m = K_n$ such that each G_i is chordal, and is obtained from its predecessor by addition of one edge (see Exercise 1.34). It is easy to see that every G_i-partial PSD matrix A_i can be "completed" to a G_{i+1}-partial PSD matrix: Suppose $e = \{p, q\}$ is the edge in G_{i+1} which is not an edge of G_i. Let K be the maximal clique in G_{i+1} containing the edge e (see Exercise 1.33) . By Proposition 1.26, the principal (partial) submatrix of A_i corresponding to K has a positive semidefinite completion. Use it to specify a value in the pq and qp positions. By induction this shows that every G-partial PSD matrix has a PSD completion. $\qquad\square$

Exercises

1.51   Show that if the diagonal entries of a partial matrix A are unspecified, and $a_{ij} = a_{ji}$ whenever both are specified, then A can be completed to a positive semidefinite matrix.

1.52   Consider the partial (but not partial symmetric) matrices

$$A = \begin{pmatrix} 2 & 2 & 2 \\ 2 & 2 & 1 \\ 2 & 1 & ? \end{pmatrix} \quad \text{and} \quad B = \begin{pmatrix} 1 & 1 & 2 \\ 1 & 2 & 2 \\ 2 & 2 & ? \end{pmatrix}.$$

Show that in both of these partial matrices, each fully specified principal submatrix is positive semidefinite, but one of them has a completion which is positive semidefinite, and the other does not.

1.53   Give an example of a C_4-partial PSD matrix which has a PSD completion.

1.54   Prove Proposition 1.26 (directly, without using Theorem 1.39).
Hint: Assume

$$A = \begin{pmatrix} d_1 & ? & \mathbf{u}^T \\ ? & d_2 & \mathbf{v}^T \\ \mathbf{u} & \mathbf{v} & E \end{pmatrix}.$$

Show that

$$\begin{pmatrix} d_1 & \mathbf{u}^T E^\dagger \mathbf{v} & \mathbf{u}^T \\ \mathbf{v}^T E^\dagger \mathbf{u} & d_2 & \mathbf{v}^T \\ \mathbf{u} & \mathbf{v} & E \end{pmatrix}$$

is a positive semidefinite completion of A.

1.55   A *G-partial PD* matrix is defined in a similar way to a G-partial PSD matrix: It is a partial symmetric matrix with specification graph G, such that each fully specified principal submatrix is positive definite. A graph G is *PD completable* if any G-partial PD matrix can be completed to a positive definite matrix. Prove that a graph G is PD completable if and only if it is PSD completable.

Notes.   The chapter is based on [Grone, Johnson, Sá and Wolkowitz (1984)]. The results in this classical paper are proved for Hermitian matrices, but we present them here in the real case.

An interesting and important result in the Grone *et al.* paper is that if a partial symmetric matrix A has a positive definite completion, then there is a unique matrix in the class of all positive definite completions of A whose determinant is maximal. This is the matrix whose inverse has zeros in the positions corresponding to the unspecified entries of A.

A partial PSD matrix A with a non-chordal specification graph may have a PSD completion (see Exercise 1.53). This depends on the numerical values of the specified entries of A. An extensive discussion of this problem is carried out in [Barrett, Johnson and Loewy (1996)] and [Barrett, Johnson and Loewy (1998)].

In the definition of a partial symmetric matrix it was required that the diagonal entries be specified. Exercises 1.51 and 1.52 give a taste of what may happen if this requirement is dropped.

Chapter 2

Complete positivity

2.1 Definition and basic properties

Consider the following question:

Question 2.1   Let $\mathbf{v}_1, \ldots, \mathbf{v}_n$ be vectors in an m-dimensional Euclidean space V. Can the vectors be embedded in a nonnegative orthant of some Euclidean space? In other words, are there a natural number k and an isometry $T : V \to \mathbf{R}^k$ such that $T\mathbf{v}_1, T\mathbf{v}_2, ..., T\mathbf{v}_n$ belong to $\mathbf{R}^k_+$?

Let A be the Gram matrix of $\mathbf{v}_1, \ldots, \mathbf{v}_n$, *i.e.*, $a_{ij} = \langle \mathbf{v}_i, \mathbf{v}_j \rangle$, $i, j = 1, \ldots, n$. Then the answer to Question 2.1 is affirmative if and only if A is (also) the Gram matrix of nonnegative vectors.

Definition 2.1   A matrix is *completely positive* if it can be decomposed as $A = BB^T$, where B is a (not necessarily square) nonnegative matrix.

The answer to Question 2.1 is "yes" if and only if $\mathrm{Gram}(\mathbf{v}_1, \ldots, \mathbf{v}_n)$ is completely positive. Here k is the number of columns of B and for each $i = 1, 2, \ldots, n$, $T\mathbf{v}_i$ is the i-th row of B.

Completely positive matrices appear in the study of block designs in combinatorics, in probability, and in various applications of statistics, including a Markovian model for DNA evolution and a model for energy demands. For more details, see the references in the notes at the end of this section.

It is very easy to construct completely positive matrices, but it may be hard to determine whether a given square matrix is completely positive or not. An obvious necessary condition for a matrix to be completely positive is that it is doubly nonnegative, that is, both positive semidefinite and entrywise nonnegative. In some cases, this condition is also sufficient, but not always. So, how can one identify completely positive matrices? This is one of the major problems in the theory of completely positive matrices, and a significant part of this chapter is devoted to it. The next chapter deals with the following (related) problem: Suppose $\mathbf{v}_1, \ldots, \mathbf{v}_n \in V$ can be embedded in the nonnegative orthant of $\mathbf{R}^m$ for some m. What is the minimal k such that $\mathbf{v}_1, \ldots, \mathbf{v}_n$ can be embedded in $\mathbf{R}_+^k$? In other words, suppose an $n \times n$ matrix A is a completely positive. What is the minimal k for which there exists a $n \times k$ nonnegative matrix B such that $A = BB^T$? In the study of these problems both the geometric approach and the algebraic approach are helpful, and we will use both.

We start with a simple example:

Example 2.1 Any positive diagonal matrix is completely positive: If $d_1, \ldots, d_n \geq 0$, then $\text{diag}(d_1, \ldots, d_n) = \text{diag}(\sqrt{d_1}, \ldots, \sqrt{d_n})^2$.

Another simple example is the matrix $J = \mathbf{e}\mathbf{e}^T$. It is also completely positive by Definition 2.1. In fact:

Example 2.2 A rank 1 doubly nonnegative matrix is necessarily completely positive: If $A \in \mathbf{R}^{n \times n}$ is a rank 1 doubly nonnegative matrix, then $A = \mathbf{b}\mathbf{b}^T$ for some $\mathbf{b} \in \mathbf{R}^n$ (by the proof of Theorem 1.10 and Remark 1.1). Since $A \geq 0$, we may take $\mathbf{b}$ to be nonnegative.

Theorem 2.1 *A rank 2 doubly nonnegative matrix is necessarily completely positive.*

Proof. Let A be an $n \times n$ doubly nonnegative matrix of rank 2. Then A is the Gram matrix of vectors $\mathbf{v}_1, \ldots, \mathbf{v}_n$ in $\mathbf{R}^2$. Since A is nonnegative, the angle between any two of these vectors is at most $\pi/2$. The two vectors which form the widest angle may therefore be rotated into the first quadrant of $\mathbf{R}^2$. In this rotation the other vectors also end up in $\mathbf{R}_+^2$. $\qquad\square$

Remark 2.1 The discussion of Question 2.1 suggests one method for showing that a doubly nonnegative matrix A is completely positive: Being positive semidefinite, A is the Gram matrix of some vectors $\mathbf{v}_1, \ldots, \mathbf{v}_n$ in some k-dimensional Euclidean space V. If an orthonormal basis E of V can be found such that the coordinate vectors $[\mathbf{v}_1]_E, [\mathbf{v}_2]_E, ..., [\mathbf{v}_n]_E$ are nonnegative, then $T\mathbf{v} = [\mathbf{v}]_E$ is an isometric embedding of $\mathbf{v}_1, \ldots, \mathbf{v}_n$ into $\mathbf{R}_+^k$, and A is completely positive. This is basically the idea in the last proof: take E to be the basis obtained by the Gram-Schmidt procedure from the two vectors which form the widest angle (that is, the pair of nonzero vectors where $\min\{\langle v_i/\|v_i\|, v_j/\|v_j\|\rangle \mid v_i, v_j \neq 0\}$ is obtained).

By Example 2.2 and Theorem 2.1, every 2×2 doubly nonnegative matrix is completely positive. In Example 2.3 we show this fact directly, this time using the algebraic approach.

Example 2.3   Let

$$A = \begin{pmatrix} a & b \\ b & c \end{pmatrix}$$

be a 2×2 doubly nonnegative matrix. Then A is completely positive, for if $c = 0$, then $A = BB^T$, where

$$B = \begin{pmatrix} \sqrt{a} & 0 \\ 0 & 0 \end{pmatrix},$$

and if $c > 0$, then $A = BB^T$ where

$$B = \begin{pmatrix} \sqrt{a - b^2/c} & b/\sqrt{c} \\ 0 & \sqrt{c} \end{pmatrix}.$$

(This decomposition may be obtained by using Schur complements — see the end of Section 1.4).

Not every doubly nonnegative matrix is completely positive.

Example 2.4   The doubly nonnegative matrix

$$A = \begin{pmatrix} 1 & 1 & 0 & 0 & 1 \\ 1 & 2 & 1 & 0 & 0 \\ 0 & 1 & 2 & 1 & 0 \\ 0 & 0 & 1 & 2 & 1 \\ 1 & 0 & 0 & 1 & 6 \end{pmatrix}$$

is not completely positive. The reader is invited to try verifying that. We will prove it in Section 2.4.

The following proposition contains an equivalent definition of complete positivity:

Proposition 2.1   *A is completely positive if and only if A can be represented as a sum*

$$A = \sum_{i=1}^{k} \mathbf{b}_i \mathbf{b}_i^T, \quad \mathbf{b}_i \geq 0, \ i = 1, \ldots, k. \tag{2.1}$$

Proof. $A = BB^T$ is equivalent to $A = \sum_{i=1}^{k} \mathbf{b}_i \mathbf{b}_i^T$, where $\mathbf{b}_i$ is the i-th column of B.   □

In (2.1) A is represented as a sum of rank 1 doubly nonnegative (completely positive) matrices.

Definition 2.2   The representation (2.1) is called a *rank 1 representation* of A.

Example 2.5   Suppose that A is as in Example 2.3, with $c = 1$. Then a rank 1 representation of A is

$$\begin{pmatrix} b^2 & b \\ b & 1 \end{pmatrix} + \begin{pmatrix} a - b^2 & 0 \\ 0 & 0 \end{pmatrix}.$$

Corollary 2.1   *The sum of completely positive matrices is completely positive.*

Proof.   Let $A = \sum_{i=1}^{k} B_i$ and $B = \sum_{i=k+1}^{k+l} B_i$ be rank 1 representations of A and B. Then $A + B = \sum_{i=1}^{k+l} B_i$.   □

The proofs of the next corollaries are similar to the corresponding proofs in Section 1.2.

Corollary 2.2   *The Hadamard product of completely positive matrices is completely positive.*

Proof.   Let $A = \sum_{i=1}^{k} \mathbf{b}_i \mathbf{b}_i^T$ and $B = \sum_{j=1}^{l} \mathbf{c}_j \mathbf{c}_j^T$ be rank 1 representations of A and B. Then

$$A \circ B = \sum_{i=1}^{k} \sum_{j=1}^{l} (\mathbf{b}_i \circ \mathbf{c}_j)(\mathbf{b}_i \circ \mathbf{c}_j)^T.$$

□

Corollary 2.3   *The Hadamard powers $A^{(k)} = A \circ A \circ \ldots \circ A$ of a completely positive matrix A are also completely positive.*

A similar result holds for ordinary powers. To prove it, we need the following proposition.

Proposition 2.2 *If A is an $n \times n$ completely positive matrix and C is an $m \times n$ nonnegative matrix, then CAC^T is also completely positive.*

Proof. Let $A = BB^T$, $B \geq 0$. Then $CAC^T = (CB)(CB)^T$. $\square$

Corollary 2.4 *If A is completely positive and k is a natural number then A^k is completely positive.*

Proof. If k is even, $k = 2l$, then $A^k = (A^l)^2 = A^l(A^l)^T$. If $k = 2l + 1$, this follows from Proposition 2.2, since $A^k = (A^l)A(A^l)^T$. $\square$

Corollary 2.5 *If $f(x)$ is a real polynomial with nonnegative coefficients and A is completely positive, then $f(A)$ is also completely positive.*

As in the case of positive semidefinite matrices, Corollary 2.2 also follows from the corresponding result for Kronecker products:

Proposition 2.3 *The Kronecker product $A \otimes C$ of completely positive matrices A and C is completely positive.*

Proof. Let $A = BB^T$, $B \geq 0$, $C = FF^T$, $F \geq 0$. The matrix $B \otimes F$ is nonnegative, and $A \otimes C = (B \otimes F)(B \otimes F)^T$. $\square$

Corollary 2.2 follows from the last proposition, since

Proposition 2.4 *Principal submatrices of completely positive matrices are completely positive.*

Proof. Let $A[\alpha]$ be a principal submatrix of a completely positive matrix A, and let $A = BB^T$ where $B \geq 0$. Then $A[\alpha]$ is a product of the submatrix of B based on the rows indexed by α and the transpose of this submatrix.

(Alternatively, if A is the Gram matrix of $\mathbf{v}_1, \ldots, \mathbf{v}_n$, then $A[\alpha]$ is the Gram matrix of $\{\mathbf{v}_i \,|\, i \in \alpha\}$. Since the whole set $\{\mathbf{v}_1, \ldots, \mathbf{v}_n\}$ may be embedded in the nonnegative orthant of some $\mathbf{R}^k$, so does any subset of it.) $\square$

The next proposition contains two simple results of Proposition 2.2, which are very useful in the study of completely positive matrices. We leave the proof as an exercise.

Proposition 2.5 *Let A be a symmetric $n \times n$ matrix. Then*

(a) *If P is an $n \times n$ permutation matrix, then A is completely positive if and only if $P^T A P$ is completely positive.*

(b) *If D is an $n \times n$ positive diagonal matrix, then A is completely positive if and only if DAD is completely positive.*

The two parts of Proposition 2.5 are algebraic statements of facts which are geometrically quite obvious: When checking whether $\mathbf{v}_1, \ldots, \mathbf{v}_n$ can be embedded in a nonnegative orthant of some $\mathbf{R}^k$, neither the order of the vectors nor their lengths matter. Here is another simple but useful observation:

Proposition 2.6   *If a symmetric matrix A is reducible, say*

$$A = A_1 \oplus A_2 \oplus \ldots \oplus A_m,$$

then A is completely positive if and only if each A_i is.

Remark 2.2   We conclude this section with a warning: The notion of complete positivity should not be confused with the notion of total positivity. However, the two positivity concepts are not totally unrelated: In Section 2.7 we will show that a symmetric totally nonnegative matrix is completely positive.

Exercises

2.1   Show that

$$A = \begin{pmatrix} 8 & 12 & 16 & 4 & 6 & 8 \\ 12 & 20 & 28 & 6 & 10 & 14 \\ 16 & 28 & 40 & 8 & 14 & 20 \\ 4 & 6 & 8 & 2 & 3 & 4 \\ 6 & 10 & 14 & 3 & 5 & 7 \\ 8 & 14 & 20 & 4 & 7 & 10 \end{pmatrix}$$

is completely positive and find a nonnegative matrix B such that $BB^T = A$.

2.2   Prove Proposition 2.5.

2.3   Prove Proposition 2.6 geometrically.

2.4   Suppose A is an $n \times n$ doubly nonnegative matrix of rank r. Then $A = \mathrm{Gram}(\mathbf{v}_1, \ldots, \mathbf{v}_n)$, where $\mathbf{v}_1, \ldots, \mathbf{v}_n \in \mathbf{R}^r$ (see Remark 1.1). Let

$\mathcal{C} = \text{cone}\,\{\mathbf{v}_1, \dots, \mathbf{v}_n\} \subseteq \mathbf{R}^r$. Show that A is completely positive if and only if there exist $\mathbf{q}_1, \dots, \mathbf{q}_m \in \mathcal{C}^*$ such that

$$\sum_{i=1}^{m} \mathbf{q}_i \mathbf{q}_i^T = I_r.$$

2.5   Let $A = \text{Gram}(\mathbf{v}_1, \dots, \mathbf{v}_n)$, where $\mathbf{v}_1, \dots, \mathbf{v}_n \in \mathbf{R}^r$. Show that if A is a rank r completely positive matrix, then $\mathcal{C} = \text{cone}\,\{\mathbf{v}_1, \dots, \mathbf{v}_n\} \subseteq \mathbf{R}^r$ is a pointed solid cone.

2.6   Let A be a doubly nonnegative matrix whose entries are all zeros and ones. Show that A is necessarily a direct sum $A_1 \oplus \dots \oplus A_k$, where each A_i is either a matrix of all ones or a zero matrix. In particular, every doubly nonnegative $(0, 1)$-matrix is completely positive.

2.7   We say that an $n \times n$ completely positive matrix A is *critical* if for every $1 \leq i \leq n$ and every $\varepsilon > 0$, $A - \varepsilon E_{ii}$ is not completely positive. Show that:

(a) Every rank 1 $n \times n$ completely positive matrix, $n \geq 2$, is critical.
(b) Every 2×2 completely positive matrix of rank 2 is not critical.

2.8   Consider the following definitions: Let A be an $n \times n$ positive semi-definite matrix, and $1 \leq i \leq n$. Let ε_0 be the maximal $\varepsilon \geq 0$ such that $A - \varepsilon E_{ii}$ is positive semidefinite. The *i-minimized matrix* of A is the matrix $A_{(i)} = A - \varepsilon_0 E_{ii}$. We say that the matrix A is *i-minimized* if $A = A_{(i)}$. A completely positive matrix A is *i-minimizable* if $A_{(i)}$ is also completely positive.

(a) Show that if

$$A = \begin{pmatrix} a & \mathbf{c}^T \\ \mathbf{c} & A_1 \end{pmatrix}$$

is positive semidefinite, then

$$A_{(1)} = \begin{pmatrix} \mu & \mathbf{c}^T \\ \mathbf{c} & A_1 \end{pmatrix},$$

where $\mu = \mathbf{c}^T A_1^\dagger \mathbf{c}$, and $A = A_{(1)}$ if and only if $\text{rank}\,A = \text{rank}\,A_1$.
(b) Every i-minimized positive semidefinite matrix is singular.
(c) Every singular positive semidefinite matrix is i-minimized for some i.

(d) If a completely positive matrix A is i-minimized for some i, then it is i-minimizable.

(e) If A is a singular completely positive matrix, then it is i-minimizable for some i.

(f) Every rank 1 completely positive matrix is i-minimizable for every i.

(g) If a completely positive matrix A is i-minimized for every i, then it is critical.

Hint: For part (a), see Theorems 1.19 and 1.20.

Notes.   The discussion of Question 2.1 is based on [Gray and Wilson (1980)] and [Hannah and Laffey (1983)]. The idea in Exercise 2.4 will be used extensively in the next chapter. It was first presented in [Hannah and Laffey (1983)]. Complete positivity of $(0,1)$-matrices (see Exercise 2.6) is discussed in [Xiang and Xiang (1998)]. The notion of a critical completely positive matrix was introduced by [Zhang and Li (2002)], and that of an i-minimizable completely positive matrix by [Barioli (1998b)]. Previous surveys of the topic of completely positive matrices are [Berman (1993)], [Ando (1991)] and [Berman (1988)].

As far as we know, the first appearance of the term "completely positive matrix" is in [Markham (1971)]. Complete positivity was first defined for quadratic forms (see [Hall and Newman (1963); Hall (1958)]):

Definition 2.3   Let A be an $n \times n$ real symmetric matrix. A real quadratic form $Q = \sum_{i,j=1}^{n} a_{ij} x_i x_j$ is called *completely positive* if it can be written as $Q = L_1^2 + L_2^2 + \ldots + L_k^2$, where $L_i = b_{i1} x_1 + b_{i2} x_2 + \ldots + b_{in} x_n$, $b_{ij} \geq 0$, $i = 1, 2, \ldots, k$, $j = 1, 2, \ldots, n$.

Clearly the quadratic form $Q = \sum_{i,j=1}^{n} a_{ij} x_i x_j$ is completely positive if and only if the matrix $A = (a_{ij})_{i,j=1}^{n}$ is completely positive.

In [Gantmacher and Krein (2002)] the name "completely positive matrices" refers to what is now called "totally nonnegative" (see Section 1.3).

The complete positivity studied in this book is not related to the concept of completely positive maps known in operator theory and used in physics (see [Benatti, Floreani and Romano (2003)]). For the sake of (positive) completeness we bring here the definition of a completely positive map (a finite dimensional version): Let M be a subspace of $\mathbf{C}^{n \times n}$ which contains the identity and such that if A belongs to M then A^*, the Hermitian adjoint of A, also belongs to M. A linear map $\varphi : M \to \mathbf{C}^{n \times n}$ is completely positive if for any positive semidefinite block matrix (X_{ij}), all of whose blocks belong to M, the block matrix $(\varphi(X_{ij}))$ is also positive semidefinite.

Completely positive matrices appear in several applications. One of the basic papers in the theory of completely positive matrices is [Gray and Wilson (1980)]. The motivation for this paper is "a proposed mathematical model of energy demand". [Drew and Johnson (1996)] mention a talk by Persi Diaconis at the Institute for Mathematics and its Applications (Minneapolis, 1993), where completely positive matrices were related to exchangeable probability distributions on finite sample spaces. Here are references to some applications in combinatorics and statistics.

Block designs. As far as we know the first application of complete positivity was to block designs. The following description is based on [Hall (1986); Hall (1958)]. A *balanced incomplete block design* $D(v, b, r, k, \lambda)$ is an arrangement of v distinct objects $a_1, \ldots, a_v$ into b blocks $B_1, \ldots, B_b$ such that each block contains exactly k distinct objects, each object occurs in exactly r different blocks, and every pair of distinct objects occurs together in exactly λ blocks. The term balanced incomplete block design comes from the theory of design of experiments in statistics. (A *complete* block design is the set of all subsets of size k of a set of v elements. A design is *balanced* if every pair of distinct elements occurs in the same number of blocks). A block design is described by its *incidence matrix* $B = (b_{ij})$, where $b_{ij} = 1$ if a_i belongs to B_j and $b_{ij} = 0$ if a_i does not belong to B_j. Each column of B describes a block and contains k 1's. Each row corresponds to an element of the set and contains r 1's. If B is the incidence matrix of $D(v, b, r, k, \lambda)$, then $A = BB^T = (r - \lambda)I_v + J_v$. Clearly A is completely positive. Complete positivity also appears in connection with the following question: Given parameters v, b, r, k, λ and blocks $B_1, \ldots, B_t$, is there a block design $D(v, b, r, k, \lambda)$ with $B_1, \ldots, B_t$ as initial blocks? Let $A = (r - \lambda)I_v + J_v$, and let $\mathbf{b}_1, \ldots, \mathbf{b}_t$ be $(0, 1)$-vectors in R^v which describe the blocks $B_1, \ldots, B_t$. Let C be a $v \times t$ matrix with columns $\mathbf{b}_1, \ldots, \mathbf{b}_t$. Then clearly a necessary condition for the existence of the block design $D(v, b, r, k, \lambda)$ with initial blocks $B_1, \ldots, B_t$ is that $A - CC^T$ is also completely positive.

Maximin efficiency-robust tests. The maximin efficiency-robust test (MERT) idea of Gastwirth (1966) is to maximize the minimum asymptotic power versus special local families of alternatives over some specially chosen families of score statistics. In the case that the number of the alternatives is finite, the MERT algorithm reduces to a maximization problem involving a correlation matrix C (a positive semidefinite matrix C with all diagonal elements equal to 1) and vectors $\mathbf{b} \in \operatorname{cs} C$ satisfying $|b_i| \geq 1$. The associated problem in which $b_i \geq 1$ has a unique solution that can be found by quadratic programming techniques. [Bose and Slud (1995)] show that if C is completely positive, then the optimal solution of the associated problem is optimal for the MERT problem. They also show that if $\rho_i \in L^2(\mathbf{R}, \mu)$, $i = 1, \ldots, m$, where μ is a Borel probability measure on the real line, and the functions ρ_i all have the same signs μ-almost everywhere, *i.e.*

$$\mu(\{t|\rho_i(t) \geq 0 \text{ for all } i, \text{ or } \rho_i(t) \leq 0 \text{ for all } i\}) = 1$$

and if c_{ij} is the inner product of ρ_i and ρ_j, then $C = (c_{ij})$ is completely positive. These hypotheses are satisfied in many versions of MERT, so in these versions the MERT problem can be solved by solving the associate problem.

Markovian model of DNA evolution. [Kelly (1994)] describes Markov models for the evolution of two species from a common ancestor. Using these models one can calculate the probability of observing two particular DNA sequences from present-day species that share a common ancestor. The data is summarized in a 4×4 matrix

$$N = \begin{pmatrix} N_{AA} & N_{AC} & N_{AG} & N_{AT} \\ N_{CA} & N_{CC} & N_{CG} & N_{CT} \\ N_{GA} & N_{GC} & N_{GG} & N_{GT} \\ N_{TA} & N_{TC} & N_{TG} & N_{TT} \end{pmatrix}$$

where A, C, G, T denote the four bases of the DNA, and N_{jk} is the number of sites in which base j is observed in the first species and base k is observed in the second. The different models are based on some simplifying assumptions and it is of interest to determine if a model assumptions are met in certain data sets. In one of the models it is assumed that the rates of evolution for all positions in the DNA sequence are equal, that the positions are independent and that the transition matrices for the two species are the same. Kelly observed that a necessary and sufficient condition for the data to fit this model is that the expected value of $N + N^T$ is completely positive and since N is a 4×4 matrix, it is enough to check that it is doubly nonnegative (See Theorem 2.4 in Section 2.3).

2.2 Cones of completely positive matrices

Let CP_n denote the set of $n \times n$ completely positive matrices. In this section we observe that CP_n is a closed convex cone, and prove that the cone CP_n and the cone COP_n of copositive matrices defined in Section 1.5 are duals of each other with respect to the inner product $\langle A, B \rangle = \text{trace } AB$ on the space S_n of symmetric $n \times n$ matrices.

Theorem 2.2 CP_n *is a closed convex cone.*

Proof. The sum of completely positive matrices is completely positive by Corollary 2.1. Let $A = BB^T$, $B \geq 0$ and let $k \geq 0$. Then

$$kA = \sqrt{k}B(\sqrt{k}B)^T,$$

so kA is also completely positive. Hence CP_n is a convex cone.

To prove that CP_n is closed, consider a sequence of completely positive matrices $\{A_i\}_{i=1}^{\infty}$ converging to A. We have to show that A is completely positive. Each of the completely positive matrices A_i is the Gram matrix of n nonnegative vectors $\mathbf{v}_1^{(i)}, \ldots, \mathbf{v}_n^{(i)}$. Let $a_{jj}^{(i)}$ denote the j-th diagonal entry of A_i. Since $\lim_{i \to \infty} \left\| \mathbf{v}_1^{(i)} \right\|^2 = \lim_{i \to \infty} a_{11}^{(i)} = a_{11}$, the sequence $\left\{ \left\| \mathbf{v}_1^{(i)} \right\| \right\}_{i=1}^{\infty}$ is bounded. The bounded sequence $\left\{ \mathbf{v}_1^{(i)} \right\}_{i=1}^{\infty}$ thus has a subsequence, $\left\{ \mathbf{v}_1^{(i_p)} \right\}_{p=1}^{\infty}$, that converges to a nonnegative vector $\mathbf{v}_1$. Consider the sequence $\left\{ \mathbf{v}_2^{(i_p)} \right\}_{p=1}^{\infty}$. Now $\lim_{p \to \infty} \left\| \mathbf{v}_2^{(i_p)} \right\|^2 = \lim_{p \to \infty} a_{22}^{(i_p)} = a_{22}$, so $\left\{ \mathbf{v}_2^{(i_p)} \right\}_{p=1}^{\infty}$ has a subsequence that converges to some $\mathbf{v}_2 \geq 0$. Continuing in this way, we finally obtain a sequence of indices $\{s\}$ such that $\lim \mathbf{v}_1^{(s)} = \mathbf{v}_1 \geq 0$, $\lim \mathbf{v}_2^{(s)} = \mathbf{v}_2 \geq 0, \ldots$, $\lim \mathbf{v}_n^{(s)} = \mathbf{v}_n \geq 0$. Thus $A = \text{Gram}(\mathbf{v}_1, \ldots, \mathbf{v}_n)$ is completely positive. $\square$

Remark 2.3 By Proposition 2.1 it is easy to see that the extreme rays of CP_n are the rays generated by rank 1 completely positive matrices — matrices of the form $\mathbf{x}\mathbf{x}^T$, where $\mathbf{x}$ is a nonnegative vector.

Theorem 2.3 CP_n *and* COP_n *are dual cones in the space* S_n. *That is,* $CP_n^* = COP_n$ *and* $COP_n^* = CP_n$

Proof. Let C be an $n \times n$ symmetric matrix. Then $C \in CP_n^*$ if and only if for every $n \times n$ completely positive matrix A, $\text{trace } (CA) \geq 0$; That is, if and only if for every nonnegative matrix B with n rows, trace $(CBB^T) \geq 0$; If and only if for every nonnegative matrix B with n rows, trace $(B^T CB) \geq 0$; If and only if for every $\mathbf{b} \in \mathbf{R}_+^n$, $\mathbf{b}^T C\mathbf{b} \geq 0$, *i.e.*, C is copositive. The dual claim that $\mathcal{COP}_n^* = \mathcal{CP}_n$ follows from Theorem 1.36, since both these cones are closed. $\square$

Remark 2.4 In Section 2.1 we observed that a necessary condition for A to be completely positive is that it is doubly nonnegative. This means that $\mathcal{DNN}_n = \mathcal{SNN}_n \cap \mathcal{PSD}_n$ contains $\mathcal{CP}_n$. Since $\mathcal{SNN}_n$ and $\mathcal{PSD}_n$ are both self dual, and in view of Theorem 1.35(e), this is consistent with the fact that $\mathcal{SNN}_n + \mathcal{PSD}_n \subseteq \mathcal{COP}_n$ (see Remark 1.10). For $n \leq 4$ $\mathcal{DNN}_n$ and $\mathcal{CP}_n$ have the same extreme rays (see Proposition 1.23 and Remark 2.3). The fact that $\mathcal{SNN}_5 + \mathcal{PSD}_5 \neq \mathcal{COP}_5$ (Example 1.30) implies that $\mathcal{DNN}_5 \neq \mathcal{CP}_5$. We will see later specific examples of 5×5 doubly nonnegative matrices which are not completely positive.

Exercises

2.9   Prove the statement in Remark 2.3.

2.10   Show that a matrix $A = (a_{ij})_{i,j=1}^n$ is completely positive if and only if there exist a measure space (Ω, μ) and nonnegative functions $f_i \in L^2(\Omega, \mu)$, $i = 1, \ldots, n$, such that

$$a_{ij} = \int_\Omega f_i(\omega) f_j(\omega) \, d\mu(\omega) \qquad i, j = 1, \ldots, n.$$

2.11   Try to find an example of a 5×5 matrix which is doubly nonnegative but not completely positive.

Notes.   The section is based on [Hall and Newman (1963)]. Exercise 2.10 is taken from [Ando (1991)], where it is used to prove a nontrivial result on Hankel matrices. A matrix $A = (a_{ij})$ is a *Hankel matrix* if $a_{i+1\,j} = a_{i\,j+1}$ for $i, j = 1, \ldots, n-1$ (the Hilbert matrices are Hankel matrices). A positive semidefinite Hankel matrix is necessarily doubly nonnegative, because of the nonnegativity of the diagonal elements. In [Ando (1991)] it is shown that if A is a positive semidefinite Hankel matrix and if the matrix obtained from A by deleting the first column and the last row is also positive semidefinite, then A is completely positive.

2.3 Small matrices.

For small matrices the necessary condition that A is doubly nonnegative is also sufficient for A to be completely positive. How small is small? At most 4×4.

Theorem 2.4 *If A is a doubly nonnegative matrix of order n, $n \leq 4$, then there exists a nonnegative matrix B such that $A = BB^T$.*

We will use the following lemma in the proof:

Lemma 2.1 *Let A and B be $n \times n$ matrices. Suppose A is completely positive, B is positive semidefinite and all its entries are zero except for a 2×2 principal submatrix, and $A + B$ is nonnegative. Then $A + B$ is completely positive.*

Proof. We may assume that the nonzero 2×2 submatrix of B is $B[1,2]$. We may also assume that B is not diagonal, otherwise B is completely positive, and so is $A + B$. So $B = \mathbf{bb}^T + \mathbf{cc}^T$, where $\mathrm{supp}\,\mathbf{b} = \{1,2\}$ and $\mathrm{supp}\,\mathbf{c} \subseteq \{1\}$ (by the same argument as in Example 2.3). Since $\mathbf{cc}^T$ is completely positive, it suffices to show that $A + \mathbf{bb}^T$ is completely positive. If $\mathbf{b} \geq \mathbf{0}$ this is clear, so we consider the case that $b_1 b_2 < 0$.

Let

$$A = \sum_{i=1}^{m} \mathbf{b}_i \mathbf{b}_i^T$$

be a rank 1 representation of A. Let

$$\mathbf{d}_i = \sqrt{\frac{(\mathbf{b}_i)_1 (\mathbf{b}_i)_2}{a_{12}}}\, \mathbf{b}\,, \quad i = 1, \ldots, m.$$

Then for each i, $\mathbf{b}_i \mathbf{b}_i^T + \mathbf{d}_i \mathbf{d}_i^T \geq 0$, since

$$
\begin{aligned}
(\mathbf{b}_i)_1 (\mathbf{b}_i)_2 + (\mathbf{d}_i)_1 (\mathbf{d}_i)_2 &= (\mathbf{b}_i)_1 (\mathbf{b}_i)_2 + \frac{(\mathbf{b}_i)_1 (\mathbf{b}_i)_2}{a_{12}} b_1 b_2 \\
&= (\mathbf{b}_i)_1 (\mathbf{b}_i)_2 \left(1 + \frac{b_1 b_2}{a_{12}} \right),
\end{aligned}
$$

and $a_{12}+b_1b_2 = (A+B)_{12} \geq 0$. Hence $\mathbf{b}_i\mathbf{b}_i^T + \mathbf{d}_i\mathbf{d}_i^T$ is a doubly nonnegative matrix of rank ≤ 2, and is therefore completely positive. Now

$$\sum_{i=1}^{m} \mathbf{d}_i\mathbf{d}_i^T = \frac{1}{a_{12}}\left(\sum_{i=1}^{m}(\mathbf{b}_i)_1(\mathbf{b}_i)_2\right)\mathbf{b}\mathbf{b}^T = \mathbf{b}\mathbf{b}^T,$$

so $A + \mathbf{b}\mathbf{b}^T = \sum_{i=1}^{m}(\mathbf{b}_i\mathbf{b}_i^T + \mathbf{d}_i\mathbf{d}_i^T)$. Hence $A + \mathbf{b}\mathbf{b}^T$ is completely positive, as a sum of m completely positive matrices.   □

Proof of Theorem 2.4.   The claim is clear for $n \leq 2$, by Examples 2.2 and 2.3.

For $n = 3$: Let A be a 3×3 doubly nonnegative matrix. If $a_{ii} = 0$, then the i-th row and the i-th column of A are equal to zero, and the result follows from the case $n \leq 2$. Thus we may assume that the diagonal entries of A are all positive. By choosing $D = \text{diag}(1/\sqrt{a_{11}}, 1/\sqrt{a_{22}}, 1/\sqrt{a_{33}})$ in Proposition 2.5(b), we see that we may assume that the diagonal entries of A are equal to 1. By Proposition 2.5(a), we may permute the rows and columns of A simultaneously. So assume that

$$A = \begin{pmatrix} 1 & a_{12} & a_{13} \\ a_{12} & 1 & a_{23} \\ a_{13} & a_{23} & 1 \end{pmatrix}$$

with

$$a_{12} \geq a_{13}. \tag{2.2}$$

We have to show that if A is doubly nonnegative, then it is completely positive. We do it by writing A as a sum of two matrices:

$$A = \mathbf{b}\mathbf{b}^T + \begin{pmatrix} & C & & 0 \\ & & & 0 \\ 0 & 0 & & 0 \end{pmatrix},$$

where

$$\mathbf{b} = \begin{pmatrix} a_{13} \\ a_{23} \\ 1 \end{pmatrix} \text{ and } C = \begin{pmatrix} 1 - a_{13}^2 & a_{12} - a_{13}a_{23} \\ a_{12} - a_{13}a_{23} & 1 - a_{23}^2 \end{pmatrix}.$$

Since $\mathbf{b}\mathbf{b}^T$ is obviously completely positive, we complete the proof by showing that C is completely positive. It is positive semidefinite, being the Schur complement $A/[3]$. It is nonnegative by (2.2), and the fact that $1 \geq a_{13}, a_{23}$

since A is positive semidefinite. Thus C is a 2×2 doubly nonnegative matrix and is therefore completely positive.

For $n = 4$: Let A be a 4×4 doubly nonnegative matrix. We have to show that A is completely positive. If $a_{ii} = 0$, then the i-th row and the i-th column of A are zero, and the problem reduces to the case $n = 3$. If the diagonal entries of A are all positive, we may assume, by Proposition 2.5, that they are all equal to 1, and that a_{34} is the smallest entry of A. Let

$$
S = \begin{pmatrix} 1 & 0 & 0 & 0 \\ 0 & 1 & 0 & 0 \\ 0 & 0 & 1 & a_{34} \\ 0 & 0 & 0 & 1 \end{pmatrix}.
$$

The matrix S is nonnegative, so by Proposition 2.2 it suffices to prove that the matrix $B = S^{-1} A (S^{-1})^T$ is completely positive.

The matrix B is equal to

$$
\begin{pmatrix}
1 & a_{12} & a_{13} - a_{14}a_{34} & a_{14} \\
a_{12} & 1 & a_{23} - a_{24}a_{34} & a_{24} \\
a_{13} - a_{14}a_{34} & a_{23} - a_{24}a_{34} & 1 - a_{34}^2 & 0 \\
a_{14} & a_{24} & 0 & 1
\end{pmatrix}.
$$

It is positive semidefinite because A is. The nonnegativity of B follows from the positive semidefiniteness of A (which implies that $a_{ij} \leq 1$ for every i, j), and the fact that a_{34} is the smallest entry of A. If $a_{34} = 1$, the problem reduces to the case $n = 3$. If $a_{34} < 1$, we can divide the third row and the third column of B by $\sqrt{1 - a_{34}^2}$ to obtain a doubly nonnegative matrix C with all diagonal entries equal to 1, and, again by Proposition 2.5, it suffices to prove that C is completely positive.

So let

$$
C = \begin{pmatrix} 1 & c_{12} & c_{13} & c_{14} \\ c_{12} & 1 & c_{23} & c_{24} \\ c_{13} & c_{23} & 1 & 0 \\ c_{14} & c_{24} & 0 & 1 \end{pmatrix}.
$$

be doubly nonnegative. To show that C is completely positive, we split C as $C = EE^T + F$, where

$$E = \begin{pmatrix} c_{13} & c_{14} \\ c_{23} & c_{24} \\ 1 & 0 \\ 0 & 1 \end{pmatrix}.$$

Then

$$F = \begin{pmatrix} f_{11} & f_{12} \\ f_{12} & f_{22} \end{pmatrix} \oplus 0_2,$$

where

$$\begin{pmatrix} f_{11} & f_{12} \\ f_{12} & f_{22} \end{pmatrix} = C/C[3,4].$$

F is therefore positive semidefinite, and EE^T completely positive. By Lemma 2.1, C is also completely positive. □

Remark 2.5 In fact, for each doubly nonnegative matrix A of order $n \leq 4$ there exists a *square* nonnegative matrix B such that $A = BB^T$. For $n \leq 3$ this follows easily from the proof of Theorem 2.4. The proof for $n = 4$ will be given in Section 3.1 (Theorem 3.3).

Theorem 2.4 can also be proved using the geometric approach. This was done for $n = 2$ in Theorem 2.1. We now demonstrate this approach in the case $n = 3$:

Geometric proof for $n = 3$. Let $\mathbf{v}_1, \mathbf{v}_2, \mathbf{v}_3$ be three vectors in $\mathbf{R}^3$ such that $\langle \mathbf{v}_i, \mathbf{v}_j \rangle \geq 0$ for $i, j = 1, 2, 3$. We may assume that all three vectors are nonzero, otherwise the problem is reduced to the case $n = 2$. We will show that there is an isometry $T : \mathbf{R}^3 \to \mathbf{R}^3$ such that $T\mathbf{v}_1, T\mathbf{v}_2$ and $T\mathbf{v}_3$ lie in $\mathbf{R}_+^3$. If the three vectors are in the same plane, then, as in the proof of Theorem 2.1, all three vectors can be embedded in $\mathbf{R}_+^2$, and thus in $\mathbf{R}_+^3$. If $\mathbf{v}_1, \mathbf{v}_2$ and $\mathbf{v}_3$ are linearly independent, let $S = \mathrm{Span}\{\mathbf{v}_1, \mathbf{v}_2\}$ and project $\mathbf{v}_3$ onto S to get $\mathbf{v}_3 = \mathbf{v}_3' + \mathbf{v}_3''$, where $\mathbf{v}_3' \in S$ and $\mathbf{v}_3'' \in S^\perp$. The vectors $\mathbf{v}_1, \mathbf{v}_2$ and $\mathbf{v}_3'$ can be embedded in $\mathbf{R}_+^2$: We may assume that the widest angle between pairs of these vectors is formed between $\mathbf{v}_1$ and one of the other vectors. Then, by applying the Gram-Schmidt procedure, we obtain an orthonormal basis $E = (\mathbf{b}_1, \mathbf{b}_2)$ of S such that the coordinate vectors $[\mathbf{v}_1]_E, [\mathbf{v}_2]_E$ and $[\mathbf{v}_3']_E$ are nonnegative, and $\mathbf{b}_1 = \mathbf{v}_1/\|\mathbf{v}_1\|$. Let

$E' = (\mathbf{b}_1, \mathbf{b}_2, \mathbf{b}_3)$, where $\mathbf{b}_3 = \mathbf{v}_3''/\|\mathbf{v}_3''\|$. Then E' is an orthonormal basis of $\mathbf{R}^3$ satisfying $[\mathbf{v}_i]_{E'} \geq 0$ for $i = 1, 2, 3$, which is what we were looking for. $\square$

Geometric proof for $n = 4$. As in the algebraic proof, the first step is to show that it is suffices to consider matrices with at least one zero entry. The geometric reasoning is this: Suppose A is a positive 4×4 doubly nonnegative matrix, and let $S = \{\mathbf{v}_1, \mathbf{v}_2, \mathbf{v}_3, \mathbf{v}_4\}$ be a set of vectors such that $A = \mathrm{Gram}(S)$. By Proposition 2.5 we may assume that $\|\mathbf{v}_i\| = 1$. We also have $0 < \langle \mathbf{v}_i, \mathbf{v}_j \rangle$, $1 \leq i, j \leq 4$. Suppose the maximal angle between the vectors is the one between $\mathbf{v}_p$ and $\mathbf{v}_q$, and project $\mathbf{v}_p$ onto $\mathrm{Span}\{\mathbf{v}_q\}$. Then $\mathbf{v}_p = \mathbf{v}_p' + \langle \mathbf{v}_p, \mathbf{v}_q \rangle \mathbf{v}_q$, where $\mathbf{v}_p'$ is orthogonal to $\mathbf{v}_q$. Let S' be the set obtained from S by replacing $\mathbf{v}_p$ by $\mathbf{v}_p'$, and let $A' = \mathrm{Gram}(S')$. If A' is completely positive, S' can be embedded in $\mathbf{R}_+^k$ for some k. And since $\mathbf{v}_p = \mathbf{v}_p' + \langle \mathbf{v}_p, \mathbf{v}_q \rangle \mathbf{v}_q$ and $\langle \mathbf{v}_p, \mathbf{v}_q \rangle \geq 0$, this implies that S can also be embedded in $\mathbf{R}_+^k$.

By Proposition 2.6, we may consider only irreducible matrices, that is, matrices whose graphs are connected. If A is a 4×4 irreducible doubly nonnegative matrix with a zero entry, its graph is one of the following:

Doubly nonnegative matrices with these graphs will be treated in Section 2.5 (algebraically and geometrically), so we skip this part of the proof. The fact that such matrices are completely positive will follow from Theorems 2.13, 2.15, 2.14, and the case $n = 3$. $\square$

Exercises

2.12 Show that if A is a 2×2 completely positive matrix, then there exist a nonnegative lower triangular matrix L such that $A = LL^T$ and a nonnegative upper triangular matrix U such that $A = UU^T$.

2.13 Let

$$A = \begin{pmatrix} 1 & a & b \\ a & 1 & c \\ b & c & 1 \end{pmatrix},$$

where $0 \le a < 1$, $0 \le b \le 1$, $ab \le c \le 1$, and $\det A \ge 0$. Check that $A = BB^T$ where

$$B = \begin{pmatrix} 1 & 0 & 0 \\ a & \sqrt{1-a^2} & 0 \\ b & \frac{c-ab}{\sqrt{1-a^2}} & \frac{\sqrt{\det A}}{\sqrt{1-a^2}} \end{pmatrix}.$$

Explain how this proves that a 3×3 matrix is completely positive if and only if it is doubly nonnegative.

2.14   Show that for $n \le 4$, an $n \times n$ completely positive matrix A with rank $A = n$ is not critical (see definition in Exercise 2.7).

Notes.   Theorem 2.4 is due to [Maxfield and Minc (1962)]. The proof here is based on [Ando (1991)], but we make use of Lemma 2.1, which is due to [Barioli (2001a)]. A geometric proof is given in [Gray and Wilson (1980)]. The geometric proofs here are taken from this paper ($n = 3$) and [Kogan (1989)] ($n = 4$). As was mentioned in Sections 1.6, the dual result, that for $n \le 4$ every $n \times n$ copositive matrix can be written as a sum of a positive semidefinite matrix and a nonnegative one is proved in [Diananda (1962)]. Exercise 2.13 is based on the original proof of Maxfield and Minc.

2.4   Complete positivity and the comparison matrix

There are two main results in this section. First, if $A \in \mathbf{R}^{n \times n}$ is nonnegative and symmetric and its comparison matrix is positive semidefinite, then A is completely positive. Second, this sufficient condition is also necessary when the graph of A contains no triangles.

We start with an easy-to-check sufficient condition for complete positivity.

Theorem 2.5   *Nonnegative symmetric diagonally dominant matrices are completely positive.*

Proof.   Let A be a nonnegative symmetric diagonally dominant matrix. Let

$$a_i = a_{ii} - \sum_{\substack{j=1 \\ j \ne i}}^{n} a_{ij},$$

then $a_i \geq 0$. Let $F_{ij} \in \mathbf{R}^{n \times n}$ have 1 in positions ii, ij, ji and jj, and 0 elsewhere. Then

$$A = \sum_{1 \leq j < i \leq n} a_{ij} F_{ij} + \text{diag}(a_1, \ldots, a_n)$$

so A is completely positive. $\square$

Example 2.6 Let

$$\begin{pmatrix} 5 & 2 & 2 \\ 2 & 4 & 1 \\ 2 & 1 & 3 \end{pmatrix}.$$

Then

$$A = \begin{pmatrix} 2 & 2 & 0 \\ 2 & 2 & 0 \\ 0 & 0 & 0 \end{pmatrix} + \begin{pmatrix} 2 & 0 & 2 \\ 0 & 0 & 0 \\ 2 & 0 & 2 \end{pmatrix} + \begin{pmatrix} 0 & 0 & 0 \\ 0 & 1 & 1 \\ 0 & 1 & 1 \end{pmatrix} + \begin{pmatrix} 1 & 0 & 0 \\ 0 & 1 & 0 \\ 0 & 0 & 0 \end{pmatrix}.$$

Hence $A = BB^T$, where

$$B = \begin{pmatrix} \sqrt{2} & \sqrt{2} & 0 & 1 & 0 \\ \sqrt{2} & 0 & 1 & 0 & 1 \\ 0 & \sqrt{2} & 1 & 0 & 0 \end{pmatrix}.$$

Theorem 2.6 *If A is symmetric and nonnegative and if its comparison matrix $M(A)$ is positive semidefinite, then A is completely positive.*

Proof. $M(A)$ is an M-matrix, so by Theorem 1.16 there exists a positive diagonal matrix D such that $DM(A)D$ is diagonally dominant. Since the entries of $DM(A)D$ and DAD are equal in absolute value, DAD is also diagonally dominant. By the previous theorem, there exists a nonnegative matrix B such that $DAD = BB^T$. But then $A = (D^{-1}B)(D^{-1}B)^T$ and since $D^{-1}B \geq 0$, A is completely positive. $\square$

Example 2.7 The matrix

$$\begin{pmatrix} 1 & 1 & 0 & 0 & 1 \\ 1 & 2 & 1 & 0 & 0 \\ 0 & 1 & 3 & 1 & 0 \\ 0 & 0 & 1 & 4 & 1 \\ 1 & 0 & 0 & 1 & 5 \end{pmatrix}$$

is completely positive, since its comparison matrix,

$$\begin{pmatrix} 1 & -1 & 0 & 0 & -1 \\ -1 & 2 & -1 & 0 & 0 \\ 0 & -1 & 3 & -1 & 0 \\ 0 & 0 & -1 & 4 & -1 \\ -1 & 0 & 0 & -1 & 5 \end{pmatrix},$$

is positive semidefinite.

The sufficient condition in Theorem 2.6 is not necessary.

Example 2.8　J_3 is completely positive but

$$M(J_3) = \begin{pmatrix} 1 & -1 & -1 \\ -1 & 1 & -1 \\ -1 & -1 & 1 \end{pmatrix}$$

is not positive semidefinite.

Observe that the graph of J_3 is a triangle. This is not a coincidence, as the following theorem shows.

Theorem 2.7　*If A is completely positive and $G(A)$ is triangle free, then $M(A)$ is positive semidefinite.*

Proof.　Let $A = \sum_{i=1}^{m} \mathbf{b}_i \mathbf{b}_i^T$ be a rank 1 representation of A. The support of each $\mathbf{b}_i$ is a clique in $G(A)$, and since $G(A)$ contains no triangles, the size of each of these supports is at most 2. Change each nonnegative vector $\mathbf{b}_i$ of support 2 to a vector $\mathbf{d}_i$ by multiplying one of its two positive entries by -1. Let $\mathbf{d}_i = \mathbf{b}_i$ if $|\text{supp}\,\mathbf{b}_i| = 1$. Then $M(A) = \sum_{i=1}^{m} \mathbf{d}_i \mathbf{d}_i^T$, so $M(A)$ is positive semidefinite. □

Combining Theorems 2.6 and 2.7 we obtain

Theorem 2.8　*If a graph G contains no triangles and A is a nonnegative symmetric matrix realization of G, then A is completely positive if and only if $M(A)$ is positive semidefinite.*

Example 2.9　The doubly nonnegative matrix

$$\begin{pmatrix} 1 & 1 & 0 & 0 & 1 \\ 1 & 2 & 1 & 0 & 0 \\ 0 & 1 & 2 & 1 & 0 \\ 0 & 0 & 1 & 2 & 1 \\ 1 & 0 & 0 & 1 & 6 \end{pmatrix}$$

is not completely positive, since its graph is the triangle free C_5, and its comparison matrix,

$$\begin{pmatrix} 1 & -1 & 0 & 0 & -1 \\ -1 & 2 & -1 & 0 & 0 \\ 0 & -1 & 2 & -1 & 0 \\ 0 & 0 & -1 & 2 & -1 \\ -1 & 0 & 0 & -1 & 6 \end{pmatrix},$$

is not positive semidefinite.

The property of G in Theorem 2.8 is, in fact, a characterization of being triangle free:

Theorem 2.9   *A graph G has the property:*

> *for every nonnegative symmetric matrix realization A of G,*
>
> *A is completely positive iff $M(A)$ is positive semidefinite*

if and only if G is triangle free.

Proof.   The "if part" is Theorem 2.8. For the "only if part", suppose G has this property, $V(G) = \{1, 2, \ldots, n\}$, and the subgraph of G induced by $\{1, 2, 3\}$ is a triangle. Define B and C in $\mathbf{R}^{n \times n}$ by

$$b_{ij} = \begin{cases} n - 1, & \text{if } i = j \\ 1, & \text{if } i \neq j \text{ and } \{i, j\} \in E(G) \\ 0, & \text{if } i \neq j \text{ and } \{i, j\} \notin E(G) \end{cases}$$

and $C = (n-1)J_3 \oplus 0_{n-3}$. The matrix C is clearly completely positive. The matrix B, being diagonally dominant, is also completely positive. Hence $A = B + C$ is completely positive. But $M(A)$ is not positive semidefinite since its 3×3 left upper principal submatrix is

$$\begin{pmatrix} 2n - 2 & -n & -n \\ -n & 2n - 2 & -n \\ -n & -n & 2n - 2 \end{pmatrix},$$

which has -2 as an eigenvalue. $\qquad\qquad\square$

The proof of Theorem 2.7 relied on the sizes of the supports of the vectors in a rank 1 representation of the matrix. It seems natural to define:

Definition 2.4 A rank 1 representation $A = \sum_{i=1}^{m} \mathbf{b}_i \mathbf{b}_i^T$, $\mathbf{b}_i \geq 0$, is *of support k* if the size of the support of each $\mathbf{b}_i$ is at most k, and at least one $\mathbf{b}_i$ has support of size exactly k.

Theorem 2.7 is therefore a special case of the following theorem, whose proof is outlined in Exercises 2.17 – 2.20.

Theorem 2.10 *Let A be a completely positive matrix which has a rank 1 representation of support k. Let $A = D + F$, where $D = \mathrm{diag}(a_{11}, \ldots, a_{nn})$. Then the matrix $M = (k-1)D - F$ is an M-matrix.*

In the case that $k = 2$ the converse is also true:

Proposition 2.7 *A completely positive matrix A has a rank 1 representation of support 2 if and only if A is nondiagonal and $M(A)$ is positive semidefinite.*

The proof of this proposition is also left as an exercise.

Exercises

2.15 Show that the following statement is false: "For every positive semidefinite matrix A, there exists a positive diagonal matrix D such that DAD is diagonally dominant".

2.16 Let A be a completely positive matrix with a triangle free graph. Let C be a real symmetric matrix with nonnegative diagonal elements such that $|C| = A$. Prove that C is positive semidefinite.

2.17 Prove this special case of Theorem 2.10: Let $B = \mathbf{b}\mathbf{b}^T$ be a rank 1 doubly nonnegative matrix. Let $B = D + F$ where $D = \mathrm{diag}(b_{11}, ..., b_{nn})$, and let $k = |\mathrm{supp}\,\mathbf{b}|$. Then $(k-1)\mathbf{e}^T D \mathbf{e} \geq \mathbf{e}^T F \mathbf{e}$.

2.18 Deduce from Exercise 2.17: Under the assumptions of Theorem 2.10, $\mathbf{e}^T M \mathbf{e} \geq 0$.

2.19 Deduce from Exercise 2.18: Under the assumptions of Theorem 2.10, $\mathbf{x}^T M \mathbf{x} \geq 0$ for every positive vector $\mathbf{x}$.

2.20 Prove: If $M = sI - B$, where $B \geq 0$ is irreducible, and $\mathbf{x}^T M \mathbf{x} \geq 0$ for every positive vector $\mathbf{x}$, then M is an M-matrix. Use this general result to complete the proof of Theorem 2.10.

2.21 Prove Proposition 2.7.

2.22   Show that the completely matrix

$$\begin{pmatrix} 2 & 1 & 0 & 0 & 1 \\ 1 & 2 & 1 & 0 & 0 \\ 0 & 1 & 2 & 1 & 0 \\ 0 & 0 & 1 & 2 & 1 \\ 1 & 0 & 0 & 1 & 2 \end{pmatrix}$$

is not i-minimizable for any $1 \leq i \leq 5$. (See definition of i-minimizability in Exercise 2.8. This example shows that the singularity requirement in part (e) of that exercise cannot be dismissed.)

Notes.   Most of the section is based on [Drew, Johnson and Loewy (1994)]. Theorem 2.5 was stated and proved in [Kaykobad (1987)] and Theorem 2.9, the proof of Theorem 2.7 and Exercise 2.16 are taken from [Berman and Shaked-Monderer (1998)]. A variation of Proposition 2.7 may be found in [Watkins (1993)].

2.5   Completely positive graphs

As mentioned in Section 2.1, an obvious necessary condition for a matrix to be completely positive is that it is doubly nonnegative. When is this necessary condition sufficient? In this section we give a qualitative answer to this question.

Definition 2.5   A graph G is *completely positive* if every doubly nonnegative matrix A whose graph is G is completely positive.

A doubly nonnegative matrix A such that $G(A) = G$ will be called a *DNN matrix realization* of the graph G. So a graph G is completely positive if every DNN matrix realization of G is completely positive. A completely positive matrix A such that $G(A) = G$ will be called a *CP matrix realization* of the graph G.

Example 2.10   Small graphs, with up to 4 vertices, are completely positive, as we saw in Section 2.3.

The main result of this section is:

Theorem 2.11   *A graph G is completely positive if and only if it does not contain an odd cycle of length greater than 4.*

We shall refer to odd cycles of length > 4 as "long odd cycles".(The triangles that played a major role in the previous section can be regarded as "small odd cycles".)

Theorem 2.11 is proved through a series of results.

Theorem 2.12　*A graph that contains a long odd cycle is not completely positive.*

Proof.　We first show that an odd cycle is not completely positive. Let B be the $k \times (k-1)$ matrix in which $b_{ii} = 1$, $i = 1, \ldots, k-1$, $b_{j+1\,j} = 1$ and $b_{kj} = (-1)^{j+1}$, $j = 1, \ldots, k-2$, and all other entries are zero:

$$B = \begin{pmatrix} 1 & 0 & \cdots & \cdots & 0 \\ 1 & 1 & \cdots & \cdots & 0 \\ 0 & 1 & 1 & \cdots & 0 \\ & & \ddots & \ddots & \\ & & & 1 & 1 \\ 1 & -1 & \cdots & \pm 1 & 0 \end{pmatrix}.$$

Let $A = BB^T$. Then

$$A = \begin{pmatrix} 1 & 1 & 0 & \cdots & & & 1 \\ 1 & 2 & 1 & 0 & & \cdots & 0 \\ 0 & 1 & 2 & 1 & & & \vdots \\ \vdots & & \ddots & \ddots & \ddots & & 0 \\ 0 & & & 1 & 2 & (-1)^{k-1} \\ 1 & 0 & \cdots & 0 & (-1)^{k-1} & k-2 \end{pmatrix}. \qquad (2.3)$$

The matrix A is positive semidefinite, and for odd k it is also nonnegative. The graph of A is a k-cycle, so for $k > 3$ it does not contain a triangle.

$$M(A) = \begin{pmatrix} 1 & -1 & 0 & \cdots & \cdots & -1 \\ -1 & 2 & -1 & 0 & \cdots & 0 \\ 0 & -1 & 2 & -1 & & \vdots \\ \vdots & & \ddots & \ddots & \ddots & 0 \\ 0 & & & -1 & 2 & -1 \\ -1 & 0 & \cdots & 0 & -1 & k-2 \end{pmatrix},$$

and $\det M(A) = k - 2 - 1 - 2 - (k-1) = -4$, so $M(A)$ is not positive semidefinite. Thus it follows from Theorem 2.7 that an odd cycle of length

$k > 4$ is not completely positive.

Now let G be a graph on n vertices, which contains a k-cycle, $k > 4$ odd. We have to show that G is not completely positive. Let S be the adjacency matrix of G and let C be an $n \times n$ matrix whose principal submatrix corresponding to the cycle is A above, and all its other entries are equal to 0. Then for every $a > 0$ there exists $0 < b \leq a$ such that $C_a = C + aI_n + bS$ is doubly nonnegative and $G(C_a) = G$. But $\lim_{a \to 0+} C_a = C$, and C is not completely positive. Since the cone $\mathcal{CP}_n$ is closed (Theorem 2.2), there exists some $a > 0$ for which C_a is doubly nonnegative but not completely positive. $\qquad\square$

Theorem 2.13 *Bipartite graphs are completely positive.*

Proof. Let A be a symmetric matrix realization of a bipartite graph G. In particular, A is permutationally similar to a block matrix

$$\begin{pmatrix} D_1 & C \\ C^T & D_2 \end{pmatrix},$$

where D_1 and D_2 are diagonal matrices.

$$\begin{pmatrix} I & 0 \\ 0 & -I \end{pmatrix} \begin{pmatrix} D_1 & C \\ C^T & D_2 \end{pmatrix} \begin{pmatrix} I & 0 \\ 0 & -I \end{pmatrix} = \begin{pmatrix} D_1 & -C \\ -C^T & D_2 \end{pmatrix}.$$

That is,

$$\begin{pmatrix} I & 0 \\ 0 & -I \end{pmatrix} A \begin{pmatrix} I & 0 \\ 0 & -I \end{pmatrix} = M(A),$$

and thus, by Sylvester's Law of Inertia (Theorem 1.7), $M(A)$ is also positive semidefinite. So, by Theorem 2.6, A is completely positive. This proves that G is completely positive. $\qquad\square$

Theorem 2.14 *Let G be a graph with a cut vertex, i.e., $G = G_1 \cup G_2$ where G_1 and G_2 have only one vertex in common. Then if both G_1 and G_2 are completely positive, so is G.*

Proof. Without loss of generality, assume that $V(G_1) = \{1, \ldots, k\}$ and $V(G_2) = \{k, k+1, \ldots, n\}$. Let A be a doubly nonnegative matrix realization of G, and suppose G_1 and G_2 are completely positive. The matrix A is the Gram matrix of some (not necessarily nonnegative) vectors $c_1, \ldots, c_n$. We have to show that there exist a Euclidean space V (not necessarily n-dimensional) and nonnegative vectors $b_1, \ldots, b_n \in V$

such that $\mathrm{Gram}(\mathbf{b}_1, \ldots, \mathbf{b}_n) = A$. The subspaces $\mathrm{Span}\{\mathbf{c}_1, \ldots, \mathbf{c}_{k-1}\}$ and $\mathrm{Span}\{\mathbf{c}_{k+1}, \ldots, \mathbf{c}_n\}$ are orthogonal. Let $\mathbf{c}_k'$ be the orthogonal projection of $\mathbf{c}_k$ on $\mathrm{Span}\{\mathbf{c}_1, \ldots, \mathbf{c}_{k-1}\}$ and let $\mathbf{c}_k''$ be the orthogonal projection of $\mathbf{c}_k$ on $\mathrm{Span}\{\mathbf{c}_{k+1}, \ldots, \mathbf{c}_n\}$. Also let $\mathbf{c}_k''' = \mathbf{c}_k - \mathbf{c}_k' - \mathbf{c}_k''$ (it may be equal to $\mathbf{0}$). Define $A' = \mathrm{Gram}(\mathbf{c}_1, \ldots, \mathbf{c}_{k-1}, \mathbf{c}_k')$. The fact that $\langle \mathbf{c}_i, \mathbf{c}_k' \rangle = \langle \mathbf{c}_i, \mathbf{c}_k \rangle$, $i = 1, \ldots, k-1$, implies that A' is equal to $A[1, \ldots, k]$, except in the lower right entry. This entry of A' is $\langle \mathbf{c}_k', \mathbf{c}_k' \rangle$, which is positive. Thus A' is doubly nonnegative, and since the graph G_1 is completely positive, there exist a Euclidean space V' and nonnegative vectors $\mathbf{d}_1, \ldots, \mathbf{d}_{k-1}, \mathbf{d}_k' \in V'$ such that $A' = \mathrm{Gram}(\mathbf{d}_1, \ldots, \mathbf{d}_{k-1}, \mathbf{d}_k')$. Similarly, for $A'' = \mathrm{Gram}(\mathbf{c}_k'', \mathbf{c}_{k+1}, \ldots, \mathbf{c}_n)$ there exists a Euclidean space V'' (of possibly different dimension) and nonnegative vectors $\mathbf{d}_k'', \mathbf{d}_{k+1}, \ldots, \mathbf{d}_n \in V''$ such that the matrix A'' is equal to $\mathrm{Gram}(\mathbf{d}_k'', \mathbf{d}_{k+1}, \ldots, \mathbf{d}_n)$. Finally, let $d_k''' = \|\mathbf{c}_k'''\|$, so that $d_k''' \geq 0$. Now let $V = V' \oplus V'' \oplus \mathbf{R}$, and let $\mathbf{b}_i = \mathbf{d}_i \oplus \mathbf{0} \oplus 0$ for $i = 1, \ldots, k-1$, $\mathbf{b}_k = \mathbf{d}_k' \oplus \mathbf{d}_k'' \oplus d_k'''$, $\mathbf{b}_i = \mathbf{0} \oplus \mathbf{d}_i \oplus 0$ for $i = k+1, \ldots, n$ be vectors in V. (The third summand in each of these vectors is redundant if $\mathbf{c}_k''' = \mathbf{0}$.) The vectors $\mathbf{b}_1, \ldots, \mathbf{b}_n$ belong to the nonnegative orthant of V, and A is their Gram matrix, which proves that A is completely positive. $\qquad \square$

Example 2.11   The "butterfly" graph

is completely positive, since 3 is a cut vertex and the triangles on vertices $\{1, 2, 3\}$ and $\{3, 4, 5\}$ are completely positive.

Theorem 2.15   *The graph T_n on n vertices, consisting of $n-2$ triangles with a common base, is completely positive.*

Proof.   Let A be a DNN matrix realization of T_n. As in the previous proofs, we may assume that A is nonsingular, and that its diagonal entries are all equal to 1. If the common base of T_n is $\{1, 2\}$, then A is of the form

$$A = \begin{pmatrix} B & C \\ C^T & I_{n-2} \end{pmatrix},$$

where B is a 2×2 doubly nonnegative matrix, and C is a positive $2 \times (n-2)$ matrix. So

$$A = \left(\begin{array}{cc} A/I_{n-2} & 0 \\ 0 & 0_{n-2} \end{array} \right) + \left(\begin{array}{c} C \\ I_{n-2} \end{array} \right) \left(\begin{array}{cc} C^T & I_{n-2} \end{array} \right).$$

By Theorem 1.19, A/I_{n-2} is positive semidefinite. Since A/I_{n-2} is 2×2, and A and C are nonnegative, A is completely positive by Lemma 2.1. $\square$

We can now prove Theorem 2.11.

Proof of Theorem 2.11. If G contains a long odd cycle, then, by Theorem 2.12, it is not completely positive. To prove the reverse implication, we show that if a graph G does not contain a long odd cycle, then every block of G is either bipartite, or the complete graph on 4 vertices, or a T_k. Thus each block is completely positive and, by Theorem 2.14, this implies that G is also completely positive. So assume that G is a block which is not bipartite and does not contain a long odd cycle. Then G must contain a triangle, say on vertices $1, 2, 3$. If G is a triangle, then $G = T_3$ and we are done. Otherwise, there exists an additional vertex in G, 4. Since G is 2-connected, 4 lies on a path P connecting two vertices of the triangle, say 1 and 2, such that no internal vertex of P is a vertex of the triangle (see Exercise 1.30). If the length of P is s, then $s \geq 2$. So G contains two cycles: $P \cup \{1, 2\}$ of length $s + 1$, and $P \cup \{1, 3\} \cup \{3, 2\}$ of length $s + 2$. If $s > 2$ then either $s + 1$ or $s + 2$ is an odd number greater than 4, hence $s = 2$.

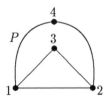

So G contains T_4. If G has more than 4 vertices, then each vertex $v \neq 1, 2, 3, 4$ lies on a path P of length s, $s \geq 2$, connecting two of the vertices $1, 2, 3, 4$. The path P cannot connect 1 and 4 (or 1 and 3, or 2 and 4, or 2 and 3), since then G would contain cycles $P \cup \{1, 2\} \cup \{2, 4\}$ of length $s + 2$ and $P \cup \{1, 3\} \cup \{3, 2\} \cup \{2, 4\}$ of length $s + 3$, and one of these cycles would be of odd length greater than 4.

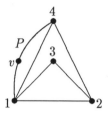

Hence P connects 1 and 2, and by the same argument as for the vertex 4, the length of P is 2. This shows that if G has n vertices, then $G \supseteq T_n$. So if $n = 4$, G is either K_4 or T_4. But if $n \geq 5$, G must be T_n, because otherwise it would have an additional edge, say $\{3, 4\}$, and this would imply that G contains a 5-cycle, $\{1, 3\} \cup \{3, 4\} \cup \{4, 2\} \cup \{2, 5\} \cup \{5, 1\}$.

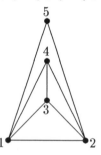

This completes the proof.   □

Graphs that do not contain long odd cycles are known to be the line perfect graphs. Thus Theorem 2.11 can be rephrased as

Corollary 2.6   *The following properties of a graph G are equivalent:*

(a) *G is completely positive.*

(b) *Each block of G is completely positive.*

(c) *Each block of G is either bipartite, or a K_4, or a T_k.*

(d) *G does not contain a long odd cycle.*

(e) *G is a line perfect graph.*

Another corollary is obtained by combining Theorems 2.11 and 2.9. It is left as an exercise:

Corollary 2.7   *A graph G has the property that*

for every nonnegative symmetric matrix realization A of G,

$M(A)$ is positive semidefinite iff A is positive semidefinite

if and only if G is bipartite.

Alternative proofs

There are other proofs of the results presented in this section. The above proofs were chosen partly for being the shortest, or the most elegant, we know. But different proofs provide different insights, so in the remainder of this section we present alternative proofs for the results of this section. We do it in the hope that the ideas in these proofs may be useful in studying other questions concerning completely positive matrices. The additional proof for the fact that bipartite graphs are completely positive (Theorem 2.13) has also the advantage of being algorithmic.

We start with the proof of Theorem 2.12. The original proof that the matrix A in (2.3) is not completely positive uses the following lemma:

Lemma 2.2 *Let A be doubly nonnegative, and let l be a vertex of $G(A)$ that does not lie on a triangle. Then A is completely positive if and only if for each neighbor j of l there exists a positive number d_j such that*

$$\sum_{a_{lj}>0} \frac{a_{lj}^2}{d_j} \le a_{ll}$$

and the matrix A' obtained from A by subtracting every d_j from a_{jj} and deleting the l-th row and column is completely positive.

Proof. We prove one direction, and leave the other as an exercise. Without loss of generality, assume that $l = 1$. Let N be the set of neighbors of 1 in $G(A)$. Suppose A is completely positive. Let

$$A = \sum_{i=1}^m \mathbf{b}_i \mathbf{b}_i^T$$

be a rank 1 representation of A. For every $j \in N$ let

$$\Omega_j = \{i \,|\, \operatorname{supp} \mathbf{b}_i = \{1, j\}\},$$

and

$$B_j = \sum_{i \in \Omega_j} \mathbf{b}_i \mathbf{b}_i^T.$$

Each B_j is zero except for its 2×2 principal submatrix based on rows and columns 1 and j. Hence by Example 2.3

$$B_j = a_j E_{11} + F_j,$$

where $a_j \geq 0$, and F_j is zero except for its $\{1, j\}$-principal submatrix, and

$$F_j[1, j] = \begin{pmatrix} a_{1j}^2/d_j & a_{1j} \\ a_{1j} & d_j \end{pmatrix}$$

for some $d_j > 0$. For these d_j's, define A' as in the statement of the lemma. Then A' is a principal submatrix of the completely positive matrix $A - \sum_{j \in N} F_j$, and is therefore completely positive.   □

We use the lemma in

Another proof of Theorem 2.12. We only demonstrate here a different way to prove that (for odd k) the doubly nonnegative matrix

$$A = \begin{pmatrix} 1 & 1 & 0 & \cdots & \cdots & 1 \\ 1 & 2 & 1 & 0 & \cdots & 0 \\ 0 & 1 & 2 & 1 & & \vdots \\ \vdots & & \ddots & \ddots & \ddots & 0 \\ 0 & & & 1 & 2 & 1 \\ 1 & 0 & \cdots & 0 & 1 & k-2 \end{pmatrix},$$

is not completely positive. The rest of the proof remains unchanged.

The graph of A is the k-cycle. None of the vertices of $G(A)$ lies on a triangle, and in particular k does not. So by the lemma, if A is completely positive then there exist positive numbers d and e such that

$$\frac{1}{d} + \frac{1}{e} \leq k - 2 \tag{2.4}$$

and the $(k-1) \times (k-1)$ matrix

$$A' = \begin{pmatrix} 1-d & 1 & 0 & \cdots & 0 \\ 1 & 2 & 1 & \cdots & 0 \\ 0 & 1 & 2 & 1 & \vdots \\ \vdots & & \ddots & \ddots & \ddots & 1 \\ 0 & & & 1 & 2-e \end{pmatrix}$$

is completely positive. But det $A'[1, 2, \ldots, k - 2] = 1 - (k - 2)d \geq 0$, hence $1/d \geq k - 2$. Since this contradicts (2.4), A is not completely positive. $\square$

Lemma 2.2 can be used to show also the following corollary, which in turn can be used to show that every tree is a completely positive graph.

Corollary 2.8 *If A is a doubly nonnegative matrix and l is a pendant vertex of $G(A)$, then A is completely positive if and only if $A/[l]$ is completely positive.*

We leave the proof of the corollary as an exercise. For the coming proofs, we point out that in order to show that G is a completely positive graph it is enough to consider only nonsingular DNN realizations of G, or consider only singular DNN realizations of G:

Lemma 2.3 *The following three conditions are equivalent:*

(a) *Every nonsingular DNN realization of G is completely positive.*
(b) *Every singular DNN realization of G is completely positive.*
(c) *G is completely positive.*

Proof. Since (c) is (a) and (b) combined, it suffices to show that (a) implies (b) and vice versa. We prove that (a)$\Rightarrow$(b), and leave the reverse implication as an exercise (Exercise 2.26).

Suppose (a) holds, and let A be a singular DNN realization of G. Then for every $\varepsilon > 0$, $A + \varepsilon I$ is also doubly nonnegative, it is nonsingular, and $G(A + \varepsilon I) = G(A)$. Now $A = \lim_{m \to \infty}(A + 1/mI)$. By the assumption, each $A + 1/mI$ is completely positive, hence by Theorem 2.2 so is A. $\square$

We now turn to the original proof of Theorem 2.13. This is a much longer proof than the one presented above. It has, however, the advantage of yielding a very simple algorithm for finding a rank 1 representation of an irreducible doubly nonnegative matrix whose graph is bipartite. Here we sketch the proof and describe the algorithm, leaving the following part of the proof as an exercise:

Proposition 2.8 *Let*

$$A = \begin{pmatrix} D_1 & C \\ C^T & D_2 \end{pmatrix},$$

where C is $k \times (n - k)$ and D_1 and D_2 are diagonal. If A is doubly nonnegative, singular and irreducible, then the null space of A is one dimensional, *and is spanned by*

$$\begin{pmatrix} \mathbf{x} \\ -\mathbf{y} \end{pmatrix},$$

(2.5)

where $\mathbf{x} \in \mathbf{R}^k$ *and* $\mathbf{y} \in \mathbf{R}^{n-k}$ *are positive.*

Another proof of Theorem 2.13. Let A be a DNN matrix realization of a bipartite graph. As usual, we may permute the rows and columns of A. So let

$$A = \begin{pmatrix} D_1 & C \\ C^T & D_2 \end{pmatrix}.$$

By Lemma 2.3 we may assume that A is singular. We may also assume that it is irreducible, for if it is reducible, it is a direct sum of irreducible matrices and the theorem can be proved for each direct summand.

The proof is by induction on the number of edges in $G(A)$. The theorem is true if $G(A)$ has one edge. Assume that it holds for graphs that have less than k edges and that $G(A)$ has k edges. By Proposition 2.8, ns A is spanned by $\mathbf{z} = (x_1, \ldots, x_k, -y_{k+1}, \ldots, -y_n)^T$, where $x_i > 0$, $i = 1, \ldots, k$ and $y_i > 0$, $i = k+1, \ldots, n$. Choose any $i < j$ such that $a_{ij} > 0$, and let $\mathbf{b}_{ij} = y_j \mathbf{e}_i + x_i \mathbf{e}_j$. This is a nonnegative vector in $\mathbf{R}^n$, orthogonal to $\mathbf{z}$. Hence $\mathbf{z}$ is in the null space of $B_{ij} = \mathbf{b}_{ij} \mathbf{b}_{ij}^T$, which implies that $\mathbf{z}$ is in the null space of every matrix of the form $A_\alpha = A - \alpha B_{ij}$. Let $\alpha_0 \geq 0$ be the largest value of α for which A_{α_0} is still doubly nonnegative. There are two possible cases: either $\alpha_0 = a_{ij}/x_i y_j$, in which case the ij entry of A_{α_0} is equal to zero, or $\alpha_0 < a_{ij}/x_i y_j$ and A_{α_0} still has positive ij and ji entries, but for $\alpha > \alpha_0$ the matrix A_α is not positive semidefinite. By the continuity of the eigenvalues and since $\mathbf{z}$ is in ns A_α for every α, the second case is, in fact, impossible. Hence $G(A_{\alpha_0})$ has fewer edges than $G(A)$, and the result follows from the induction hypothesis. $\square$

As mentioned above, this proof suggests a simple algorithm. If A_{α_0} is still irreducible, its null space is spanned by $\mathbf{z}$ and we may repeat the procedure to eliminate another edge from the graph. If A_{α_0} is reducible, then it is permutationally similar to a direct sum of irreducible matrices. In this case a permutation of $\mathbf{z}$ will be a null vector for the direct sum and since $\mathbf{z}$ has no zero entries, each of the summands is singular. The nullity of each summand is then exactly 1, and we can proceed to eliminate another edge using the newly labeled version of G. At each stage we obtain a matrix that is permutationally similar to a direct sum of doubly nonnegative singular matrices, so when we finish eliminating the off-diagonal entries the diagonal entries also vanish, since 1×1 singular matrices are equal to zero. Observe that the permutation of rows and columns at each stage so that each irreducible summand has the special form

$$\left(\begin{array}{cc} D_1 & C \\ C^T & D_2 \end{array} \right)$$

was needed only to simplify the description of $\mathbf{z}$, but is not really necessary in the actual factorization. The algorithm is therefore simpler:

Let A be an irreducible doubly nonnegative matrix whose graph is bipartite.

1. Obtain from A a row equivalent upper triangular matrix U by forward elimination (that is, using only operations of subtracting from one row scalar multiples of preceding rows — this is possible since all the leading principal minors, except possibly $\det A$ itself, are positive).

2. Let $U_1 = U - u_{nn}E_{nn}$. Find a vector $\mathbf{z}$ that spans ns U_1.

3. For every $i < j$ such that $a_{ij} > 0$: If $z_j \mathbf{e}_i - z_i \mathbf{e}_j \geq \mathbf{0}$, let $b_{ij} = z_j \mathbf{e}_i - z_i \mathbf{e}_j$, otherwise let $b_{ij} = z_i \mathbf{e}_j - z_j \mathbf{e}_i$.

4.

$$A = u_{nn}E_{nn} + \sum_{\substack{i<j \\ a_{ij}>0}} \frac{a_{ij}}{-z_i z_j} \mathbf{b}_{ij} \mathbf{b}_{ij}^T.$$

For $n \times n$ matrices, the above algorithm requires about $n^3/3$ multiplications.

Algebraic proof of Theorem 2.14. Let $G = G_1 \cup G_2$, where G_1 and G_2 are completely positive graphs which share exactly one vertex. Suppose that $V(G_1) = \{1, 2, \ldots, k\}$, $V(G_2) = \{k, k+1, \ldots, n\}$. To prove that G is completely positive, we have to show that every DNN matrix realization A of G is completely positive. Suppose then that A is a doubly nonnegative matrix of the block form

$$A = \left(\begin{array}{cc} A_{11} & A_{21}^T \\ A_{21} & A_{22} \end{array} \right),$$

where A_{11} is a $k \times k$ matrix and A_{21} is an $(n-k) \times k$ matrix in a block form $\left(\begin{array}{cc} 0_{(n-k)\times(k-1)} & \mathbf{y} \end{array} \right)$. By Theorem 1.19 the (generalized) Schur complement A/A_{22} is positive semidefinite. Since the first $k-1$ columns in A_{21} are zero,

$$A/A_{22} = A_{11} - \mathbf{y}^T A_{22}^\dagger \mathbf{y} E_{kk},$$

where E_{kk} is a $k \times k$ matrix whose last right entry is 1 and all other entries are 0. So the entries of A/A_{22} are the same as the entries of A_{11} except for the kk entry which is positive, being a diagonal entry of a positive definite matrix. Thus A/A_{22} is doubly nonnegative and has the same graph as A_{11}, which is the completely positive G_1. Hence A/A_{22} is completely positive. Consider now the $(n-k+1) \times (n-k+1)$ matrix

$$A' = \left(\begin{array}{cc} \mathbf{y}^T A_{22}^\dagger \mathbf{y} & \mathbf{y}^T \\ \mathbf{y} & A_{22} \end{array} \right).$$

Since A_{22} is positive semidefinite, so is $A_{22}^\dagger$, and $\mathbf{y}^T A_{22}^\dagger \mathbf{y} \geq 0$. The matrix A' is also positive semidefinite by Theorem 1.19, so it is doubly nonnegative. $G(A') = G_2$, so A' is completely positive. Finally,

$$A = \left(\begin{array}{cc} A/A_{22} & 0 \\ 0^T & 0_{n-k} \end{array} \right) + \left(\begin{array}{cc} 0_{k-1} & 0 \\ 0 & A' \end{array} \right)$$

so A is completely positive.   □

Geometric proof of Theorem 2.15.   The proof is by induction on n. For $n = 3$ the graph is a triangle, which is a small graph and is thus completely positive by Theorem 2.4. Suppose that every T_k, $3 \leq k < n$, is completely positive, and consider T_n. Let A be a DNN matrix realization of T_n. The matrix A is the Gram matrix of vectors $\mathbf{c}_1, \ldots, \mathbf{c}_n$. We want to show that $\mathbf{c}_1, \ldots, \mathbf{c}_n$ may be embedded in some m-dimensional vector space V (m may be different than n), with an orthonormal basis E such that $[\mathbf{c}_1]_E, [\mathbf{c}_2]_E, \ldots, [\mathbf{c}_n]_E$ belong to $\mathbf{R}_+^m$. We may assume that $\{1,2\}$ is the common base of T_n, and since complete positivity is not affected by positive diagonal congruence, we may assume that $a_{ii} = 1$ for every $i = 1, \ldots, n$. By this assumption, $\|\mathbf{c}_i\| = 1$ and $\langle \mathbf{c}_i, \mathbf{c}_j \rangle \leq 1$, $1 \leq i \neq j \leq n$. Let $\langle \mathbf{c}_p, \mathbf{c}_q \rangle$ be the minimal inner product of all the positive $\langle \mathbf{c}_i, \mathbf{c}_j \rangle$'s. Consider two cases:

Case 1. $\{p, q\} = \{1, 2\}$. Define $\mathbf{c}_1' = \mathbf{c}_1 - \langle \mathbf{c}_1, \mathbf{c}_2 \rangle \mathbf{c}_2$, and consider the vectors $\mathbf{c}_1', \mathbf{c}_2, \ldots, \mathbf{c}_n$. Then $\langle \mathbf{c}_1', \mathbf{c}_2 \rangle = 0$, and for $i > 2$,

$$\langle \mathbf{c}_1', \mathbf{c}_i \rangle = \langle \mathbf{c}_1, \mathbf{c}_i \rangle - \langle \mathbf{c}_1, \mathbf{c}_2 \rangle \langle \mathbf{c}_2, \mathbf{c}_i \rangle \geq \langle \mathbf{c}_1, \mathbf{c}_i \rangle - \langle \mathbf{c}_1, \mathbf{c}_2 \rangle \geq 0.$$

This shows that the positive semidefinite matrix $A' = \text{Gram}(c_1', c_2, \ldots, c_n)$ is doubly nonnegative. The graph of A' is obtained from $G(A)$ by erasing the base (and possibly other edges), hence it is bipartite, and as such completely positive. Hence we may assume that $c_1', c_2, \ldots, c_n \in V$, for some vector space V with an orthonormal basis E such that $[c_1']_E, [c_2]_E, \ldots, [c_n]_E$ are nonnegative. But $[c_1]_E = [c_1']_E + \langle c_1, c_2 \rangle [c_2]_E$, so $[c_1]_E$ is also nonnegative.

Case 2. $\{p, q\} \neq \{1, 2\}$. Without loss of generality we may assume that $\{p, q\} = \{1, n\}$. Defining $c_1' = c_1 - \langle c_1, c_n \rangle c_n$, we get (as in Case 1) vectors $\{c_1', c_2, \ldots, c_n\}$ with a nonnegative Gram matrix A', whose graph is obtained from $G(A)$ by erasing the edge $\{1, n\}$ and possibly some other edges incident with the vertex 1. The vertex 2 is a cut vertex of the graph $G(A')$, connecting a graph T_k with $\{1, 2\}$ for a base and $k < n$, the edge $\{2, n\}$, and possibly some other edges.

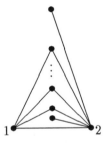

By the induction hypothesis and Theorem 2.14, $G(A')$ is completely positive and so is A'. Thus there is a vector space V with an orthonormal basis E such that $[c_1']_E, [c_2]_E, \ldots, [c_n]_E$ are nonnegative. But $c_1 = c_1' + \langle c_1, c_n \rangle c_n$, so $[c_1]_E$ is also nonnegative and A is completely positive. $\square$

Exercises

2.23 Let

$$A = \begin{pmatrix} 1 & a & b & 0 & 0 \\ a & 1 & c & 0 & 0 \\ b & c & 1 & d & e \\ 0 & 0 & d & 1 & f \\ 0 & 0 & e & f & 1 \end{pmatrix}$$

be doubly nonnegative. Find a nonnegative matrix B such that $BB^T = A$.

2.24   Complete the proof of Lemma 2.2.

2.25   Prove Corollary 2.8. Then use the corollary to prove by induction that every tree is a completely positive graph.

2.26   Let A be a nonsingular DNN realization of a graph G. Use Proposition 1.10 to prove that $A - (\det A/\det A(n))E_{nn}$ is a singular DNN realization of G. Use that to prove that in Lemma 2.3 (b)$\Rightarrow$(a).

2.27   Prove Proposition 2.8, using the following steps:

(a) Show that if Q is an $k \times (n - k)$ matrix,

$$S = \begin{pmatrix} 0 & Q \\ Q^T & 0 \end{pmatrix}$$

is nonnegative and irreducible, and $\mathbf{x} \in \mathbf{R}^k$ and $\mathbf{y} \in \mathbf{R}^{n-k}$, then

$$S\begin{pmatrix} \mathbf{x} \\ \mathbf{y} \end{pmatrix} = \lambda \begin{pmatrix} \mathbf{x} \\ \mathbf{y} \end{pmatrix} \text{ iff } S\begin{pmatrix} \mathbf{x} \\ -\mathbf{y} \end{pmatrix} = -\lambda \begin{pmatrix} \mathbf{x} \\ -\mathbf{y} \end{pmatrix}.$$

(b) Show that if S is as in (a), then the smallest eigenvalue of S is simple and the associated eigenspace is spanned by a vector of the form (2.5).

(c) Prove Proposition 2.8.

2.28   Use the algorithm that follows from the original proof of Theorem 2.13 to find a rank 1 representation of

$$A = \begin{pmatrix} 6 & 0 & 2 & 2 \\ 0 & 5 & 4 & 2 \\ 2 & 4 & 6 & 0 \\ 2 & 2 & 0 & 6 \end{pmatrix}.$$

2.29   By Proposition 2.8, If A is an $n \times n$ completely positive matrix whose graph is bipartite, then rank $A \geq n - 1$. Show that every bipartite graph has both a nonsingular CP matrix realization and a singular one.

2.30   Let G be a completely positive graph. Let A be a completely positive matrix with $G(A) = G$. Show that

(a) A is critical if and only if $A - \varepsilon E_{ii}$ is not positive semidefinite for every $1 \leq i \leq n$.

(b) If A is critical, then A is singular.

(c) A is i-minimizable for every i.

2.31 Show that if A is a doubly nonnegative matrix realization of an odd cycle, then A is completely positive if and only if $\det M(A) \geq 0$.

Notes. Theorem 2.11 is due to [Kogan and Berman (1993)]. See also [Ando (1991)], [Berman (1988)], [Berman (1993)], [Kogan (1989)] and [Drew and Johnson (1996)] (where graphs which contain no long odd cycles are called *NLOC graphs*). Theorem 2.12 is due to [Berman and Hershkowitz (1987)]. The first part of the proof of this theorem, which appears on page 84, is taken from [Berman and Shaked-Monderer (1998)]. The second part is taken from the original proof, and the rest of the original proof is given on page 90. The deletion of a pendant vertex in Corollary 2.8 is used in [Berman and Hershkowitz (1987)] to show that trees are completely positive. This idea is extended in Section 2.6 to a deletion of a vertex of degree 2. Theorem 2.13 is due to [Berman and Grone (1988)]. The original proof is given on page 92. Proposition 2.8 appears in [Berman and Grone (1988)], and the proof outlined in Exercise 2.27 is based on that paper. The equation in part (a) of that exercise, for the case that S is the adjacency matrix of a bipartite graph, is known in chemistry as the Coulson-Rushbrooke Pairing Theorem. Theorem 2.14 is proved in [Kogan and Berman (1993)]. The algebraic proof (page 93) is taken from [Ando (1991)]. The graphs T_n discussed in this section are called *crowns* in [Salce (1993)]. The geometric proof of Theorem 2.15 (page 94) is the original proof of [Kogan and Berman (1993)]. The proof that we give on page 86 is a simplification of the algebraic proof of [Ando (1991)], using Lemma 2.1 (which is due to [Barioli (2001a)]). Corollary 2.7 is taken from [Berman and Shaked-Monderer (1998)]. Exercise 2.30 is due to [Zhang and Li (2002)] (parts (a) and (b)) and to [Barioli (1998b)] (part (c)). Exercise 2.31 is taken from [Zhang and Li (2000)].

2.6 Completely positive matrices whose graphs are not completely positive

In this section we collect miscellaneous results on complete positivity of matrices whose graphs are not completely positive. The section is divided into three subsections: positivity of least squares solution, book graphs and odd cycles, and deletion of vertices of degree 2.

Positivity of least squares solution

We describe an inductive sufficient condition for complete positivity. Consider an $n \times n$ matrix A in block form

$$A = \begin{pmatrix} a & \mathbf{c}^T \\ \mathbf{c} & A_1 \end{pmatrix} \tag{2.6}$$

where A_1 is a symmetric $(n-1) \times (n-1)$ matrix.

Theorem 2.16 *Let A be an $n \times n$ matrix in the block form (2.6). Then A is completely positive if and only if A_1 is completely positive and there exist a (not necessarily square) nonnegative matrix C and a nonnegative vector $\mathbf{x}$ such that $A_1 = CC^T$, $\mathbf{c} = C\mathbf{x}$, and $a = \mathbf{x}^T\mathbf{x}$.*

Proof. If A is completely positive, then $A = BB^T$, where B is nonnegative. Denote the first row of B by $\mathbf{x}^T$, and let C be the matrix consisting of rows $2, \ldots, n$ of B. Then $\mathbf{x}$ and C are nonnegative and satisfy the required equalities, since

$$BB^T = \begin{pmatrix} \mathbf{x}^T \\ C \end{pmatrix} \begin{pmatrix} \mathbf{x} & C^T \end{pmatrix} = \begin{pmatrix} \mathbf{x}^T\mathbf{x} & \mathbf{x}^T C^T \\ C\mathbf{x} & CC^T \end{pmatrix}.$$

Conversely, suppose there exist C and $\mathbf{x}$ satisfying the conditions in the theorem. Then

$$B = \begin{pmatrix} \mathbf{x}^T \\ C \end{pmatrix}$$

is a nonnegative matrix satisfying $A = BB^T$. □

The matrix C and the vector $\mathbf{x}$ in the statement of Theorem 2.16 are not unique.

Definition 2.6 A symmetric nonnegative matrix A in block form (2.6) has *property PLSS (positivity of least squares solution)* if there exists a nonnegative matrix C such that $CC^T = A_1$ and $C^\dagger \mathbf{c}$ is a nonnegative vector.

If A_1 is completely positive and the first row of A is zero, then A trivially has property PLSS. Here is another example:

Example 2.12 If A is a nonnegative symmetric matrix in block form (2.6) and A_1 is a nonnegative rank 1 symmetric matrix, then A has property

PLSS. To see that, note that such A_1 is equal to $\mathbf{u}\mathbf{u}^T$ for some nonnegative vector $\mathbf{u}$. The matrix A has the PLSS property since $\mathbf{u}^\dagger = \|\mathbf{u}\|^{-2}\mathbf{u}^T$.

The next example is left as an exercise.

Example 2.13 If A is a nonnegative symmetric matrix in block form (2.6) and A_1 is a diagonal matrix, then A has property PLSS.

Observe that if A has property PLSS, then so does every nonnegative matrix which differs from A only in its 11 entry.

Theorem 2.17 *A doubly nonnegative matrix that has the PLSS property is completely positive.*

Proof. Let $C \geq 0$ satisfy $CC^T = A_1$ and $\mathbf{x} = C^\dagger \mathbf{c} \geq 0$. Then the matrix

$$\begin{pmatrix} \mathbf{x}^T \\ C \end{pmatrix} \begin{pmatrix} \mathbf{x} & C^T \end{pmatrix} = \begin{pmatrix} \mathbf{x}^T\mathbf{x} & \mathbf{x}^TC^T \\ C\mathbf{x} & CC^T \end{pmatrix} = \begin{pmatrix} \mathbf{x}^T\mathbf{x} & \mathbf{c}^T \\ \mathbf{c} & CC^T \end{pmatrix}$$

is completely positive. By Theorem 1.19,

$$a - \mathbf{x}^T\mathbf{x} = a - \mathbf{c}^T(C^\dagger)^T C^\dagger \mathbf{c} = a - \mathbf{c}^T A_1^\dagger \mathbf{c} \geq 0.$$

Hence

$$A = \begin{pmatrix} a - \mathbf{x}^T\mathbf{x} & \mathbf{0}^T \\ \mathbf{0} & 0 \end{pmatrix} + \begin{pmatrix} \mathbf{x}^T\mathbf{x} & \mathbf{x}^TC^T \\ C\mathbf{x} & CC^T \end{pmatrix}$$

is completely positive. $\qquad\qquad\square$

The converse of Theorem 2.17 is not true.

Example 2.14 Let

$$A = \begin{pmatrix} 2 & 1 & 0 & 0 & 1 \\ 1 & 2 & 1 & 0 & 0 \\ 0 & 1 & 2 & 1 & 0 \\ 0 & 0 & 1 & 2 & 1 \\ 1 & 0 & 0 & 1 & 2 \end{pmatrix}.$$

The matrix A is completely positive since it is diagonally dominant. However, A does not have property PLSS. If it had that property, then so would its 1-minimized matrix

$$
\begin{pmatrix}
1.2 & 1 & 0 & 0 & 1 \\
1 & 2 & 1 & 0 & 0 \\
0 & 1 & 2 & 1 & 0 \\
0 & 0 & 1 & 2 & 1 \\
1 & 0 & 0 & 1 & 2
\end{pmatrix},
$$

but the latter is not completely positive since its comparison matrix is not positive semidefinite.

Unlike double nonnegativity or complete positivity, the PLSS property is not preserved under permutation similarity.

Example 2.15   Let A be as in Example 2.14, B the 6×6 matrix in block form

$$
B = \begin{pmatrix} 0 & \mathbf{0}^T \\ \mathbf{0} & A \end{pmatrix}
$$

and

$$
B' = \begin{pmatrix} A & \mathbf{0} \\ \mathbf{0}^T & 0 \end{pmatrix}.
$$

The matrices B and B' are permutationally similar, B trivially has the PLSS property, but B' does not, since A does not have this property.

Since property PLSS is not preserved under permutation similarity, it is worthwhile to mention the following obvious corollary of Theorem 2.17:

Corollary 2.9   *If PAP^T has property PLSS for some permutation matrix P, then A is completely positive.*

The matrix A in Example 2.14 can be used to show that the converse of Corollary 2.9 also does not hold. We leave the proof as an exercise, together with the proof of the following proposition, which asserts that some permutation similarities do preserve the PLSS property.

Proposition 2.9   *If $P = 1 \oplus P'$, where P' is an $(n-1) \times (n-1)$ permutation matrix, then A has property PLSS if and only if PAP^T has this property.*

The following is a corollary of Theorem 2.17 and Examples 2.12 and 2.13.

Corollary 2.10 *If A is a doubly nonnegative matrix in block form* (2.6) *and* rank $A_1 = 1$ *or* A_1 *is diagonal, then A is completely positive.*

Property PLSS can be used to verify complete positivity. We leave it as an exercise to show that $n \times n$ doubly nonnegative matrices, $n \leq 4$, are not only completely positive — they actually have the PLSS property. It is also possible to prove directly that a matrix whose graph is T_n, where 1 is one of the vertices of the common base, has property PLSS, thus providing an alternative proof to Theorem 2.15.

Book graphs and odd cycles

Definition 2.7 A graph G is a *book graph* with r completely positive pages if $V(G) = V_1 \cup \ldots \cup V_r$, where

(a) There exist two distinct vertices u and v such that $V_i \cap V_j = \{u, v\}$ for every $1 \leq i \neq j \leq r$.
(b) The subgraph of G induced by V_i, G_i, is completely positive.
(c) $G = G_1 \cup \ldots \cup G_r$.

Note that $\{u, v\}$ may or may not be an edge of G, and that by condition (c) two vertices of G which belong to different V_i's are not adjacent, unless they are u and v and $\{u, v\} \in E(G)$. Several examples of book graphs are shown bellow.

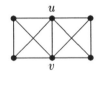 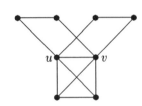 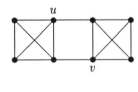

A matrix realization of a book graph is permutationally similar to a matrix A in the following block form

$$
A = \begin{pmatrix}
a & h & \mathbf{x}_1^T & \mathbf{x}_2^T & \cdots & \mathbf{x}_r^T \\
h & b & \mathbf{y}_1^T & \mathbf{y}_2^T & \cdots & \mathbf{y}_r^T \\
\mathbf{x}_1 & \mathbf{y}_1 & A_1 & 0 & \cdots & 0 \\
\mathbf{x}_2 & \mathbf{y}_2 & 0 & A_2 & \cdots & 0 \\
\vdots & \vdots & \vdots & \vdots & \ddots & \vdots \\
\mathbf{x}_r & \mathbf{y}_r & 0 & 0 & \cdots & A_r
\end{pmatrix} \tag{2.7}
$$

where for every $1 \leq i \leq r$ the graph of

$$\begin{pmatrix} a & h & \mathbf{x}_i^T \\ h & b & \mathbf{y}_i^T \\ \mathbf{x}_i & \mathbf{y}_i & A_i \end{pmatrix}$$

is the completely positive G_i. Let $\tilde{A}$ be the direct sum of $A_1, \ldots, A_r$, $\mathbf{x}^T = (\mathbf{x}_1^T, \mathbf{x}_2^T, \ldots, \mathbf{x}_r^T)$ and $\mathbf{y}^T = (\mathbf{y}_1^T, \mathbf{y}_2^T, \ldots, \mathbf{y}_r^T)$. Then

$$A = \begin{pmatrix} a & h & \mathbf{x}^T \\ h & b & \mathbf{y}^T \\ \mathbf{x} & \mathbf{y} & \tilde{A} \end{pmatrix}. \tag{2.8}$$

Denote

$$\Delta = \{1 \leq i \leq r \mid \mathbf{x}_i^T A_i^\dagger \mathbf{y}_i < 0\} \text{ and } \Delta' = \{1, \ldots, r\} \setminus \Delta.$$

Using these notations we state a necessary and sufficient condition for a DNN matrix realization of a book graph to be completely positive.

Theorem 2.18   *Let A be a doubly nonnegative matrix such that $G(A)$ is a book graph, A in block form (2.7), (2.8). Then A is completely positive if and only if $\Delta = \emptyset$, or $\Delta \neq \emptyset$ and*

$$\left(a - \mathbf{x}^T \tilde{A}^\dagger \mathbf{x}\right)^{1/2} \left(b - \mathbf{y}^T \tilde{A}^\dagger \mathbf{y}\right)^{1/2} \geq -2 \sum_{i \in \Delta} \mathbf{x}_i^T A_i^\dagger \mathbf{y}_i. \tag{2.9}$$

For the proof we need the following lemma:

Lemma 2.4   *Let*

$$A = \begin{pmatrix} a & h \\ h & b \end{pmatrix}$$

be a real positive semidefinite 2×2 matrix and $y \in \mathbf{R}$. There exist $x, z \in \mathbf{R}$ such that

$$\begin{pmatrix} x & y \\ y & z \end{pmatrix} \text{ and } \begin{pmatrix} a - x & h - y \\ h - y & b - z \end{pmatrix}$$

are both positive semidefinite if and only if

$$\frac{h - \sqrt{ab}}{2} \leq y \leq \frac{h + \sqrt{ab}}{2}. \tag{2.10}$$

Proof. If

$$\begin{pmatrix} x & y \\ y & z \end{pmatrix} \text{ and } \begin{pmatrix} a-x & h-y \\ h-y & b-z \end{pmatrix}$$

are positive semidefinite, then so are

$$\begin{pmatrix} x & -y \\ -y & z \end{pmatrix}$$

and

$$\begin{pmatrix} a & h-2y \\ h-2y & b \end{pmatrix} = \begin{pmatrix} a-x & h-y \\ h-y & b-z \end{pmatrix} + \begin{pmatrix} x & -y \\ -y & z \end{pmatrix}.$$

The fact that the matrix on the left is positive semidefinite implies that $ab \geq (h-2y)^2$, and the desired inequalities follow.

The converse is clearly true if either $a = 0$ or $b = 0$. If $ab > 0$, then $h = \sqrt{ab}\cos 2\theta$ for some $0 \leq \theta \leq \pi/2$. So $A = \cos^2\theta A_1 + \sin^2\theta A_2$, where

$$A_1 = \begin{pmatrix} a & \sqrt{ab} \\ \sqrt{ab} & b \end{pmatrix} \text{ and } A_2 = \begin{pmatrix} a & -\sqrt{ab} \\ -\sqrt{ab} & b \end{pmatrix}.$$

The matrices A_1 and A_2 are both positive semidefinite.

By (2.10),

$$-\sin^2\theta \leq \frac{y}{\sqrt{ab}} \leq \cos^2\theta.$$

If $0 \leq (y/\sqrt{ab}) \leq \cos^2\theta$, we may choose

$$\begin{pmatrix} x & y \\ y & z \end{pmatrix} = \frac{y}{\sqrt{ab}}A_1,$$

$$\begin{pmatrix} a-x & h-y \\ h-y & b-z \end{pmatrix} = \left(\cos^2\theta - \frac{y}{\sqrt{ab}}\right)A_1 + \sin^2\theta A_2.$$

If $-\sin^2\theta \leq (y/\sqrt{ab}) \leq 0$, we may choose

$$\begin{pmatrix} x & y \\ y & z \end{pmatrix} = \frac{-y}{\sqrt{ab}}A_1,$$

$$\begin{pmatrix} a-x & h-y \\ h-y & b-z \end{pmatrix} = \left(\sin^2\theta + \frac{y}{\sqrt{ab}}\right)A_1 + \cos^2\theta A_2.$$

$\square$

In the proof of Theorem 2.18 we use a few more notations. For each $1 \le i \le r$, let

$$\begin{pmatrix} a_i & h_i \\ h_i & b_i \end{pmatrix} = \begin{pmatrix} \mathbf{x}_i^T \\ \mathbf{y}_i^T \end{pmatrix} A_i^\dagger \begin{pmatrix} \mathbf{x}_i & \mathbf{y}_i \end{pmatrix}$$

and let C_i be the $n \times n$ matrix with zero entries except for $C_i[V_i]$, which is

$$\begin{pmatrix} a_i & h_i & \mathbf{x}_i^T \\ h_i & b_i & \mathbf{y}_i^T \\ \mathbf{x}_i & \mathbf{y}_i & A_i \end{pmatrix}.$$

Let

$$\begin{pmatrix} a_\perp & h_\perp \\ h_\perp & b_\perp \end{pmatrix} = A/\tilde{A}.$$

By Theorem 1.19 the last matrix is positive semidefinite, and since

$$\begin{pmatrix} \mathbf{x}^T \\ \mathbf{y}^T \end{pmatrix} \tilde{A}^\dagger \begin{pmatrix} \mathbf{x} & \mathbf{y} \end{pmatrix} = \sum_{i=1}^{r} \begin{pmatrix} \mathbf{x}_i^T \\ \mathbf{y}_i^T \end{pmatrix} A_i^\dagger \begin{pmatrix} \mathbf{x}_i & \mathbf{y}_i \end{pmatrix}$$

$$= \sum_{i=1}^{r} \begin{pmatrix} a_i & h_i \\ h_i & b_i \end{pmatrix},$$

$A = \sum_{i=1}^{r} C_i + C_\perp$, where

$$C_\perp = \begin{pmatrix} a_\perp & h_\perp \\ h_\perp & b_\perp \end{pmatrix} \oplus 0_{n-2}.$$

With these notations, $h_\perp + \sum_{i=1}^{r} h_i = h$, and condition (2.9) becomes $\sqrt{a_\perp b_\perp} + h_\perp \ge -2\sum_{i \in \Delta} h_i$.

Proof of Theorem 2.18. First we show that if A is completely positive, then condition (2.9) holds. Let

$$A = \sum_{j=1}^{m} \mathbf{b}_j \mathbf{b}_j^T$$

be a rank 1 representation of A. For $1 \le j \le r - 1$ let

$$\Omega_i = \{1 \le j \le m \mid \operatorname{supp} \mathbf{b}_j \cap (V_i \setminus \{1, 2\}) \ne \emptyset\},$$

and let

$$\Omega_r = \{1, \ldots, m\} \setminus \cup_{i=1}^{r-1} \Omega_i.$$

Finally for $1 \leq i \leq r$ let

$$B_i = \sum_{j \in \Omega_i} \mathbf{b}_j \mathbf{b}_j^T .$$

Each B_i is a completely positive matrix and B_i is zero except for $B_i[V_i]$ which is

$$\begin{pmatrix} d_i & e_i & \mathbf{x}_i^T \\ e_i & f_i & \mathbf{y}_i^T \\ \mathbf{x}_i & \mathbf{y}_i & A_i \end{pmatrix} .$$

By Theorem 1.19,

$$B_i[V_i]/A_i = \begin{pmatrix} d_i & e_i \\ e_i & f_i \end{pmatrix} - \begin{pmatrix} a_i & h_i \\ h_i & b_i \end{pmatrix}$$

is positive semidefinite for every $1 \leq i \leq r$. Hence

$$\begin{pmatrix} a_\Delta & h_\Delta \\ h_\Delta & b_\Delta \end{pmatrix} = \sum_{i \in \Delta} \left(\begin{pmatrix} d_i & e_i \\ e_i & f_i \end{pmatrix} - \begin{pmatrix} a_i & h_i \\ h_i & b_i \end{pmatrix} \right)$$

and

$$\begin{pmatrix} a_{\Delta'} & h_{\Delta'} \\ h_{\Delta'} & b_{\Delta'} \end{pmatrix} = \sum_{i \in \Delta'} \left(\begin{pmatrix} d_i & e_i \\ e_i & f_i \end{pmatrix} - \begin{pmatrix} a_i & h_i \\ h_i & b_i \end{pmatrix} \right)$$

are both positive semidefinite. Since

$$\sum_{i=1}^{r} B_i = A = \sum_{i=1}^{r} C_i + C_\perp,$$

the equality

$$\sum_{i=1}^{r} \begin{pmatrix} d_i & e_i \\ e_i & f_i \end{pmatrix} = \sum_{i=1}^{r} \begin{pmatrix} a_i & h_i \\ h_i & b_i \end{pmatrix} + \begin{pmatrix} a_\perp & h_\perp \\ h_\perp & b_\perp \end{pmatrix}$$

holds. Hence

$$\begin{pmatrix} a_\Delta & h_\Delta \\ h_\Delta & b_\Delta \end{pmatrix} + \begin{pmatrix} a_{\Delta'} & h_{\Delta'} \\ h_{\Delta'} & b_{\Delta'} \end{pmatrix} = \begin{pmatrix} a_\perp & h_\perp \\ h_\perp & b_\perp \end{pmatrix}$$

By Lemma 2.4 this implies that $\sqrt{a_\perp b_\perp} + h_\perp \geq 2h_\Delta$, which in turn implies (2.9), since $h_\Delta \geq -\sum_{i \in \Delta} h_i \geq 0$.

For the converse, we consider three possible cases:

Case 1. $\Delta = \emptyset$ and $h_\perp \geq 0$. In this case, $C_\perp$ is doubly nonnegative and therefore completely positive. Every C_i is doubly nonnegative and $G(C_i)$ is the completely positive graph G_i, hence every C_i is also completely positive. So A is completely positive as a sum of $r+1$ completely positive matrices.

Case 2. $\Delta = \emptyset$ and $h_\perp < 0$. In this case $\sum_{i=1}^r h_i > 0$ and we may define

$$C_i' = C_i + \frac{h_i}{\sum_{i=1}^r h_i} C_\perp.$$

It is easy to see that each C_i' is doubly nonnegative and $G(C_i') = G_i$. Hence it is completely positive and so is $A = \sum_{i=1}^r C_i'$.

Case 3. $\Delta \neq \emptyset$ and $\sqrt{a_\perp b_\perp} + h_\perp \geq -2\sum_{i\in\Delta} h_i$. In this case, we may apply Lemma 2.4 to

$$\begin{pmatrix} a_\perp & h_\perp \\ h_\perp & b_\perp \end{pmatrix} \quad \text{and} \quad y = -\sum_{i\in\Delta} h_i$$

and get $x, z \in \mathbf{R}$ such that

$$\begin{pmatrix} x & y \\ y & z \end{pmatrix} \quad \text{and} \quad \begin{pmatrix} a_\perp - x & h_\perp - y \\ h_\perp - y & b_\perp - z \end{pmatrix}$$

are both positive semidefinite. Let

$$C_i' = \begin{cases} C_i - \dfrac{h_i}{y} \begin{pmatrix} x & y & \mathbf{0}^T \\ y & z & \mathbf{0}^T \\ \mathbf{0} & \mathbf{0} & 0_{n-2} \end{pmatrix} & i \in \Delta \\[3em] C_i + \dfrac{h_i}{\sum_{j\in\Delta'} h_j} \begin{pmatrix} a_\perp - x & h_\perp - y & \mathbf{0}^T \\ h_\perp - y & b_\perp - z & \mathbf{0}^T \\ \mathbf{0} & \mathbf{0} & 0_{n-2} \end{pmatrix} & i \in \Delta' \end{cases}$$

Each C_i' is a doubly nonnegative matrices with $G(C_i') \subseteq G_i$, and is therefore completely positive. Hence $A = \sum_{i=1}^r C_i'$ is also completely positive. $\square$

For the statement of the following corollary, we denote

$$A_0 = \begin{pmatrix} a & h \\ h & b \end{pmatrix} \tag{2.11}$$

and for $1 \leq i \leq r$

$$H_i^T = \begin{pmatrix} \mathbf{x}_i^T \\ \mathbf{y}_i^T \end{pmatrix}. \tag{2.12}$$

The proof of the corollary is left as an exercise.

Corollary 2.11   *Let A be a doubly nonnegative matrix such that $G(A)$ is a book graph. Suppose A is in block form (2.7), A_0 and H_i are defined as in (2.11) and (2.12). Then A is completely positive if and only if the matrix $A_0 - \sum_{i=1}^r \left| H_i^T A_i^\dagger H_i \right|$ is positive semidefinite.*

Next we consider a special type of book graph.

Definition 2.8   A graph G is a $\widehat{CP}$ *graph* if there exists a vertex $v \in V(G)$ such that $d(v) = 2$, the two vertices adjacent to v are not adjacent to each other, and $G - v$ is completely positive.

So a $\widehat{CP}$ graph is created by adding a vertex v to a completely positive graph, along with two edges connecting it to two nonadjacent vertices u and w. Here are two examples:

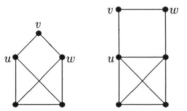

If G is a $\widehat{CP}$ graph on n vertices, we may assume that $v = 1$ and the two nonadjacent vertices which are neighbors of v are $u = 2$ and $w = n$. A symmetric matrix realization of G then has this form:

$$A = \begin{pmatrix} a & \mathbf{c}^T \\ \mathbf{c} & \tilde{A} \end{pmatrix} \tag{2.13}$$

where $\mathbf{c} = \mathbf{c}_1 + \mathbf{c}_2 \in \mathbf{R}^{n-1}$,

$$\mathbf{c}_1^T = (\gamma_2, 0, \ldots, 0) \quad \text{and} \quad \mathbf{c}_2^T = (0, \ldots, 0, \gamma_n),$$

and $a_{2n} = a_{n2} = 0$.

We denote $\overline{\mathbf{c}} = \mathbf{c}_1 - \mathbf{c}_2$ and

$$A^- = \begin{pmatrix} a & \overline{\mathbf{c}}^T \\ \overline{\mathbf{c}} & \tilde{A} \end{pmatrix}. \tag{2.14}$$

Using these notations we now state a necessary and sufficient condition for a matrix whose graph is $\widehat{CP}$ to be completely positive.

Theorem 2.19 *Let A be a doubly nonnegative matrix whose graph is $\widehat{\text{CP}}$. Suppose A be in block form (2.13) and let A^- be as in (2.14). Then A is completely positive if and only if A^- is positive semidefinite.*

Proof. If A is completely positive, let $A = \sum_{i=1}^{m} \mathbf{b}_i \mathbf{b}_i^T$ be a rank 1 representation of A, and let

$$\Omega_1 = \{1 \leq i \leq m \mid \{1, n\} = \text{supp } \mathbf{b}_i\} \text{ and } \Omega_2 = \{1, \ldots, m\} \setminus \Omega_1.$$

For every $i \in \Omega_1$ let $\mathbf{d}_i$ be the vector obtained from $\mathbf{b}_i$ by multiplying the first entry by -1. Then

$$A^- = \sum_{i \in \Omega_1} \mathbf{d}_i \mathbf{d}_i^T + \sum_{i \in \Omega_2} \mathbf{b}_i \mathbf{b}_i^T$$

and is therefore positive semidefinite.

We now prove the sufficiency of this condition. Suppose A and A^- are both positive semidefinite. Then there exist $n \times n$ matrices W and $\overline{W}$ such that $A = WW^T$ and $A^- = \overline{W}\,\overline{W}^T$. Let

$$W = \begin{pmatrix} \mathbf{w}^T \\ W_0 \end{pmatrix} \text{ and } \overline{W} = \begin{pmatrix} \overline{\mathbf{w}}^T \\ \overline{W}_0 \end{pmatrix}.$$

Since $W_0 W_0^T = \overline{W}_0 \overline{W}_0^T$, there exists an orthogonal $n \times n$ matrix U such that $U W_0^T = \overline{W}_0^T$, and we may replace $\overline{W}$ by $\overline{W} U^T$. So assume

$$W = \begin{pmatrix} \mathbf{w}^T \\ W_0 \end{pmatrix} \text{ and } \overline{W} = \begin{pmatrix} \overline{\mathbf{w}}^T \\ W_0 \end{pmatrix}.$$

Then

$$A = \begin{pmatrix} \mathbf{w}^T \mathbf{w} & \mathbf{w}^T W_0^T \\ W_0 \mathbf{w} & W_0 W_0^T \end{pmatrix} \text{ and } A^- = \begin{pmatrix} \overline{\mathbf{w}}^T \overline{\mathbf{w}} & \overline{\mathbf{w}}^T W_0^T \\ W_0 \overline{\mathbf{w}} & W_0 W_0^T \end{pmatrix},$$

and in particular

$$W_0 \mathbf{w} = \mathbf{c} \text{ and } W_0 \overline{\mathbf{w}} = \overline{\mathbf{c}}. \tag{2.15}$$

Let

$$\mathbf{u} = \frac{1}{2}(\mathbf{w} + \overline{\mathbf{w}}) \text{ and } \mathbf{v} = \frac{1}{2}(\mathbf{w} - \overline{\mathbf{w}}).$$

Obviously, $\mathbf{w} = \mathbf{u} + \mathbf{v}$ and since $\|\mathbf{w}\|^2 = \|\overline{\mathbf{w}}\|^2 = a$, $\mathbf{u}$ and $\mathbf{v}$ are orthogonal. By (2.15),

$$W_0 \mathbf{u} = \frac{1}{2}(\mathbf{c} + \overline{\mathbf{c}}) = \mathbf{c}_1 \text{ and } W_0 \mathbf{v} = \frac{1}{2}(\mathbf{c} - \overline{\mathbf{c}}) = \mathbf{c}_2.$$

Let

$$\hat{A} = \begin{pmatrix} \mathbf{u} \\ \mathbf{v} \\ W_0 \end{pmatrix} \begin{pmatrix} \mathbf{u} \\ \mathbf{v} \\ W_0 \end{pmatrix}^T.$$

Then $\hat{A}$ is an $(n+1) \times (n+1)$ positive semidefinite matrix, and since $\mathbf{u}$ and $\mathbf{v}$ are orthogonal,

$$\hat{A} = \begin{pmatrix} \mathbf{u}^T \mathbf{u} & 0 & \mathbf{c}_1^T \\ 0 & \mathbf{v}^T \mathbf{v} & \mathbf{c}_2^T \\ \mathbf{c}_1 & \mathbf{c}_2 & \tilde{A} \end{pmatrix}.$$

Thus the matrix $\hat{A}$ is nonnegative, and its graph is the graph of $\tilde{A}$ with two additional vertices, each joined by an edge to a vertex of $G(\tilde{A})$. The graph $G(\hat{A})$ is therefore completely positive, and so is the matrix $\hat{A}$. Let

$$E = \begin{pmatrix} 1 & 1 & 0 & 0 & \ldots & 0 \\ 0 & 0 & 1 & 0 & \ldots & 0 \\ 0 & 0 & 0 & 1 & \ldots & 0 \\ \vdots & \vdots & \vdots & & \ddots & \vdots \\ 0 & 0 & 0 & \ldots & 0 & 1 \end{pmatrix} \in \mathbf{R}^{n \times (n+1)}.$$

Then $A = E\hat{A}E^T$. This proves that A is completely positive. $\square$

Remark 2.6 Observe that the proof of the sufficiency part in the last theorem did not rely on the fact that $a_{2n} = a_{n2} = 0$. So if a graph G has a vertex v such that $d(v) = 2$ and $G - v$ is completely positive, and A is a doubly nonnegative realization of G, we may assume that A is in block form (2.13) and define A^- as in (2.14). If A^- is positive semidefinite, then A is completely positive.

Odd cycles of length ≥ 5 are $\widehat{\mathrm{CP}}$ graphs which are not completely positive. We use Theorem 2.19 to obtain a qualitative necessary and sufficient condition for a DNN matrix realization of such a graph to be completely positive. If A is a doubly nonnegative matrix whose graph is the n-cycle, we may assume that A has this form:

$$
\begin{pmatrix}
a_1 & h_1 & 0 & 0 & & \cdots & 0 & h_n \\
h_1 & a_2 & h_2 & 0 & & \cdots & 0 & 0 \\
0 & h_2 & a_3 & h_3 & & \cdots & 0 & 0 \\
0 & 0 & h_3 & a_4 & \ddots & & \vdots & \vdots \\
\vdots & \vdots & & \ddots & \ddots & \ddots & & \vdots \\
0 & 0 & \cdots & & \ddots & a_{n-2} & h_{n-2} & 0 \\
0 & 0 & \cdots & \cdots & & h_{n-2} & a_{n-1} & h_{n-1} \\
h_n & 0 & \cdots & \cdots & & 0 & h_{n-1} & a_n
\end{pmatrix}
\tag{2.16}
$$

Theorem 2.20 *Let A be a doubly nonnegative matrix realization of the n-cycle, $n \geq 5$ odd. Suppose A is in the form (2.16). Then A is completely positive if and only if*

$$
\det A \geq 4 \prod_{i=1}^{n} h_i .
\tag{2.17}
$$

Proof. By Theorem 2.19 A is completely positive if and only if A^- is positive semidefinite. But since n is odd,

$$
\det A - \det A^- = 4h_1 h_n \det
\begin{pmatrix}
h_2 & a_3 & h_3 & \cdots & 0 & 0 \\
0 & h_3 & a_4 & \cdots & 0 & 0 \\
\vdots & \vdots & \vdots & \ddots & \vdots & \vdots \\
0 & 0 & 0 & \cdots & a_{n-2} & h_{n-2} \\
0 & 0 & 0 & \cdots & h_{n-2} & a_{n-1} \\
0 & 0 & 0 & \cdots & 0 & h_{n-1}
\end{pmatrix}
$$

$$
= 4 \prod_{i=1}^{n} h_i .
$$

Hence $\det A \geq 4 \prod_{i=1}^{n} h_i$ if and only if $\det A^- \geq 0$. If A is completely positive, then A^- is positive semidefinite, and therefore inequality (2.17) holds. Conversely, if (2.17) holds, then $\det A > 0$, so the doubly nonnegative A is nonsingular. But then $A^-(1) = A(1)$ is positive definite, and therefore $\det A^- \geq 0$ implies that A^- is positive semidefinite. $\qquad\square$

Remark 2.7 Theorem 2.20 implies that a completely positive matrix whose graph is an odd cycle of length 5 or more is necessarily positive definite.

Deletion of a vertex of degree 2

In this subsection we extend the idea of deletion of a pendant vertex (Corollary 2.8) to deletion of a vertex of degree 2. Suppose then that a graph G has a vertex v such that $d(v) = 2$ and u and w are the vertices adjacent to v. Given a doubly nonnegative matrix A whose graph is G, we construct a smaller doubly nonnegative matrix A' whose graph is obtained from G be deletion of the vertex v and possibly also the edge $\{u, w\}$. We may assume that $V(G) = \{1, \ldots, n\}$, $v = 1$, $u = 2$ and $w = 3$. Then if A is a DNN matrix realization of G,

$$A = \begin{pmatrix} a_{11} & a_{12} & a_{13} & 0 & \ldots & 0 \\ a_{12} & a_{22} & a_{23} & \ldots & \ldots & a_{2n} \\ a_{13} & a_{23} & a_{33} & & & \vdots \\ 0 & \vdots & & \ddots & & \\ \vdots & \vdots & & & \ddots & \vdots \\ 0 & a_{2n} & \ldots & & \ldots & a_{nn} \end{pmatrix}. \tag{2.18}$$

We write

$$A = \begin{pmatrix} \tilde{A} & S^T \\ S & M^\bullet \end{pmatrix} \tag{2.19}$$

where $M = A[3, \ldots, n]$, and define

$$A^S = \begin{pmatrix} S^T M^\dagger S & S^T \\ S & M \end{pmatrix}$$

and

$$A^R = A - A^S = (A/M) \oplus 0_{n-2}.$$

By Theorem 1.19 the matrices A^S and A^R are positive semidefinite. Let

$$A_t = tA^S + (1-t)A.$$

For every $0 \leq t \leq 1$ the matrix A_t is positive semidefinite, $A^S \preceq A_t \preceq A$, and $A_t[3, \ldots, n] = M$. This, together with the fact that $A - A_t$ is positive semidefinite, implies that the last $n - 2$ rows and columns of A_t are equal to those of A. In particular, most of the entries of A_t are nonnegative — the 12 entry is the only one which may be negative.

Let

$$\delta(A) = \det A[1, 2|1, 3] = a_{11}a_{23} - a_{12}a_{13}.$$

For every t,

$$\delta(A_t) = t\delta(A^S) + (1 - t)\delta(A) = t(\delta(A^S) - \delta(A)) + \delta(A). \qquad (2.20)$$

Now let

$$\alpha = \begin{cases} 0 & \text{if } \delta(A) \geq 0 \\ \dfrac{-\delta(A)}{\delta(A^S) - \delta(A)} & \text{if } \delta(A) < 0 \leq \delta(A^S) \\ 1 & \text{otherwise} \end{cases}$$

and

$$A^{[1/2]} = A_\alpha = A - \alpha A^R.$$

That is, if $\delta(A) \geq 0$, $A^{[1/2]} = A$; If $\delta(A) < 0 \leq \delta(A^S)$, $A^{[1/2]} = A_t$ for the unique t such that $\delta(A_t) = 0$; and if both $\delta(A)$ and $\delta(A^S)$ are negative, $A^{[1/2]} = A^S$. Finally, let

$$A^{[1]} = A^{[1/2]}/[1].$$

Remark 2.8 The graph $G(A^{[1]})$ is obtained from $G(A)$ by deleting the vertex 1 and possibly also the edge $\{2, 3\}$.

Lemma 2.5 *Let A be a doubly nonnegative matrix in the form (2.18), then $A^{[1/2]}$ is also doubly nonnegative, and $A^{[1]}$ is positive semidefinite. The matrix $A^{[1]}$ is doubly nonnegative if and only if $\max(\delta(A), \delta(A^S)) \geq 0$.*

Proof. To show that $A^{[1/2]}$ is doubly nonnegative we only need to show that its 12 entry is nonnegative. This is clear if $\delta(A) \geq 0$, since in that case $A^{[1/2]} = A$. If $\delta(A) < 0 \leq \delta(A^S)$, then $\delta(A^{[1/2]}) = 0$, so that

$$(A^{[1/2]})_{12} = \frac{(A^{[1/2]})_{11}(A^{[1/2]})_{23}}{(A^{[1/2]})_{13}} \geq 0.$$

Finally, if $\delta(A)$ and $\delta(A^S)$ are both negative, then $\delta(A^{[1/2]}) = \delta(A^S) < 0$, and this implies that

$$(A^{[1/2]})_{12} > \frac{(A^{[1/2]})_{11}(A^{[1/2]})_{23}}{(A^{[1/2]})_{13}} \geq 0.$$

The matrix $A^{[1]}$ is positive semidefinite by Theorem 1.19, and only its $1\,2$ entry may be negative. But

$$(A^{[1]})_{12} = (A^{[1/2]})_{23} - \frac{(A^{[1/2]})_{12}(A^{[1/2]})_{13}}{(A^{[1/2]})_{11}} = \frac{\delta(A^{[1/2]})}{(A^{[1/2]})_{11}},$$

so $A^{[1]}$ is nonnegative if and only if $\delta(A^{[1/2]}) \geq 0$. By the definition of $A^{[1/2]}$ and (2.20), this occurs if and only if $\max(\delta(A), \delta(A^S)) \geq 0$. $\square$

Remark 2.9   If $A^{[1]}$ is completely positive, then so is $A^{[1/2]}$, since

$$A^{[1/2]} - (0_1 \oplus A^{[1]}) = \frac{1}{(A^{[1/2]})_{11}} \mathbf{u}\mathbf{u}^T,$$

where $\mathbf{u}$ is the first column of $A^{[1/2]}$.

If $A^{[1/2]}$ is completely positive, then by Lemma 2.1 A is also completely positive, since $A - A^{[1/2]}$ is a positive semidefinite matrix which is zero except for a 2×2 principal submatrix.

So if $A^{[1]}$ is completely positive, A is also completely positive.

If $\delta(A) \geq 0$, then the converse is also true:

Theorem 2.21   *Let A be a doubly nonnegative matrix in the form* (2.18) *with $\delta(A) \geq 0$. Then A is completely positive if and only if $A^{[1]}$ is completely positive.*

Proof.   Suppose A is completely positive. Let $A = \sum_{i=1}^{m} \mathbf{b}_i \mathbf{b}_i^T$ be any rank 1 representation of A, and let

$$\Omega_1 = \{1 \leq i \leq m \,|\, 1 \in \operatorname{supp} \mathbf{b}_i\} \text{ and } \Omega_2 = \{1, \ldots, m\} \setminus \Omega_1.$$

Let

$$B = \sum_{i \in \Omega_1} \mathbf{b}_i \mathbf{b}_i^T \text{ and } C = \sum_{i \in \Omega_2} \mathbf{b}_i \mathbf{b}_i^T.$$

Then B and C are completely positive matrices and $A = B + C$. The matrix B is zero except for its 3×3 leading principal submatrix, so the matrix $H = 0_1 \oplus B/[1]$ is zero except for its principal submatrix based on rows 2 and 3.

Observe that

$$A - (0_1 \oplus A/[1]) = B - H = \begin{pmatrix} a_{11} & a_{12} & a_{13} \\ a_{12} & a_{12}^2/a_{11} & a_{12}a_{13}/a_{11} \\ a_{13} & a_{12}a_{13}/a_{11} & a_{13}^2/a_{11} \end{pmatrix} \oplus 0_{n-3}.$$

We may therefore write

$$0_1 \oplus A^{[1]} = A - (B - H) = C + H.$$

Since C is completely positive and $C+H$ is nonnegative, $C+H$ is completely positive by Lemma 2.1. Hence $0_1 \oplus A^{[1]}$ is completely positive, and so is $A^{[1]}$. $\square$

If $\delta(A) < 0$, it is possible for A to be completely positive when $A^{[1]}$ is not.

Example 2.16 Let

$$A = \begin{pmatrix} 2 & 1 & 1 & 0 & 0 \\ 1 & 2 & 0 & 0 & 1 \\ 1 & 0 & 2 & 1 & 0 \\ 0 & 0 & 1 & 2 & 1 \\ 0 & 1 & 0 & 1 & 2 \end{pmatrix}.$$

Then A is diagonally dominant and therefore completely positive, and

$$A^S = \begin{pmatrix} 5/4 & 3/4 & 1 & 0 & 0 \\ 3/4 & 5/4 & 0 & 0 & 1 \\ 1 & 0 & 2 & 1 & 0 \\ 0 & 0 & 1 & 2 & 1 \\ 0 & 1 & 0 & 1 & 2 \end{pmatrix}.$$

Hence in this case $\max(\delta(A), \delta(A^S)) < 0$, and $A^{[1]}$ is not even doubly nonnegative.

In the case that $\delta(A) < 0$, an additional condition is required to ensure that complete positivity of A is equivalent to that of $A^{[1]}$.

Definition 2.9 Let A be a doubly nonnegative matrix in the form (2.19) and let 1 be a vertex of degree 2 in $G(A)$. We say that A is *diminishable* if rank $A \leq$ rank $M + 1$.

Remark 2.10 Clearly, every diminishable matrix is singular. The converse is generally not true. But if M is nonsingular and A is singular then A is diminishable.

The diminishability of A implies that A_t, $0 \leq t \leq 1$, are the only matrices satisfying $A^S \preceq B \preceq A$:

Lemma 2.6 *Let A be a diminishable doubly nonnegative matrix in block form (2.19). Then every matrix B such that $A^S \preceq B \preceq A$ is a convex combination of A^S and A.*

Proof. By Theorem 1.20, rank $A = $ rank $A^R + $ rank M. Since A is diminishable, this implies that rank $A^R \leq 1$. But then if $A^S \preceq B \preceq A$, the equality

$$A^R = (A - B) + (B - A^S)$$

implies that each of the positive semidefinite matrices $A - B$ and $B - A^S$ is a nonnegative multiple of A^R. $\qquad \square$

We can now prove:

Theorem 2.22 *If A is a doubly nonnegative diminishable matrix of the form (2.18), then A is completely positive if and only if $A^{[1]}$ is completely positive.*

Proof. We already know that if $A^{[1]}$ is completely positive, then so are $A^{[1/2]}$ and A (see Remark 2.9). By Theorem 2.21 we only need to show that if A is completely positive and $\delta(A) < 0$ then $A^{[1]}$ is completely positive. We first show that under these conditions $A^{[1/2]}$ is completely positive.

So suppose A is completely positive. Let $A = \sum_{i=1}^m \mathbf{b}_i \mathbf{b}_i^T$ be any rank 1 representation of A, and let

$$\Omega_1 = \{1 \leq i \leq m \mid 1 \in \text{supp } \mathbf{b}_i\} \text{ and } \Omega_2 = \{1, \ldots, m\} \setminus \Omega_1.$$

Let

$$B = \sum_{i \in \Omega_1} \mathbf{b}_i \mathbf{b}_i^T \text{ and } C = \sum_{i \in \Omega_2} \mathbf{b}_i \mathbf{b}_i^T.$$

Then $B = B' \oplus 0_{n-3}$ and $C = 0_1 \oplus C'$ are completely positive matrices such that $A = B + C$, and

$$\delta(B') = b_{11}b_{23} - b_{12}b_{13} = a_{11}b_{23} - a_{12}a_{13} \leq \delta(A) < 0.$$

Hence

$$b_{12}b_{13} > b_{11}b_{23}.$$

Multiply this inequality by the inequality

$$b_{11}b_{33} \geq b_{13}^2$$

to get that

$$b_{33}b_{12} > b_{13}b_{23}. \tag{2.21}$$

Inequality (2.21) implies that $H = B'/[3] \oplus 0_{n-2}$ is nonnegative. The matrix $B - H$ is equal to $\mathbf{u}\mathbf{u}^T$ where $\mathbf{u}$ is the third column of B divided by $\sqrt{b_{33}}$, so $\mathbf{u}$ is a nonnegative vector with supp $\mathbf{u} = \{1, 2, 3\}$. Hence $B - H$ is also completely positive, and $F = A - H = C + \mathbf{u}\mathbf{u}^T$ is completely positive. Since $F[3, \ldots, n] = M$, the matrix F^S is equal to A^S, which implies that $A^S \preceq F$. Since H is positive semidefinite, we also have $F \preceq A$. By Lemma 2.6, $F = A_\beta$ for some $0 \leq \beta \leq 1$. Now

$$\delta(A_\beta) = f_{11}f_{23} - f_{12}f_{13} = u_1^2(c_{23} + u_2u_3) - u_1^2 u_2 u_3 = u_1^2 c_{23} \geq 0. \tag{2.22}$$

Combining (2.20) and (2.22) and the fact that $\delta(A) < 0$, we get that $\delta(A^S) \geq 0$. But then (again by (2.20)) $\delta(A_t)$ is an increasing function. Hence $\delta(A^{[1/2]}) = 0$ implies that $\alpha \leq \beta$. We may write

$$A^{[1/2]} = A_\beta + (\beta - \alpha)A^R.$$

So $A^{[1/2]}$ is completely positive, as a sum of a completely positive matrix and a positive semidefinite matrix which is zero except for a 2×2 principal submatrix. Since $\delta(A^{[1/2]}) = 0$, Theorem 2.21 implies that $(A^{[1/2]})^{[1]} = A^{[1]}$ is also completely positive. □

Remark 2.11   Observe that Theorem 2.22 also holds when 1 is a vertex of degree 1. In that case $\delta(A) = 0$, and $A^{[1]} = A/[1]$. By Corollary 2.8, the theorem holds in this case too.

Successive deletion of pendant vertices may be used to show that every DNN matrix realization of a tree is completely positive. A similar use of the technic of deletion of a vertex of degree two is possible, if we start with a diminishable matrix. To be able to use the technic on $A^{[1]}$, the matrix $A^{[1]}$ needs to be diminishable with respect to some vertex of degree 2 in

$G(A^{[1]})$, or $G(A^{[1]})$ has to have a vertex of degree 1. We describe a family of graphs for which such repeated deletions are possible.

Definition 2.10 We say that a graph G on n vertices is an *NCC graph* if G has two vertices of degree 2, u and v, and $E(G)$ consists of two internally disjoint paths P and Q from u to v,

$$P = \{u, p_1\}, \{p_1, p_2\}, \ldots, \{p_{m-2}, p_{m-1}\}, \{p_{m-1}, v\}$$

$$Q = \{v, q_1\}, \{q_1, q_2\}, \ldots, \{q_{k-2}, q_{k-1}\}, \{q_{k-1}, v\},$$

and edges $\{p_i, q_j\}$ such that:

If for $i \leq l$ both $\{p_i, q_j\}$ and $\{p_l, q_r\}$ are edges of G, then $j \leq r$.   (2.23)

The two paths form a cycle, and the additional edges are noncrossing chords of the cycle, with one end on one path and another end on the other path. Here is one example of an NCC graph:

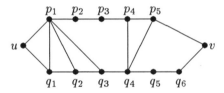

While every NCC graph is a cycle with noncrossing chords, not every cycle with noncrossing chords is an NCC graph. For example,

is not an NCC graph.

Given an NCC graph G on n vertices as in Definition 2.10, we order its vertices. For that we define $\nu : V(G) \to \{1, \ldots, n\}$ as follows: $\nu(u) = 1$ and $\nu(v) = n$. Next, we consider p_1 and q_1. If there exists $l > 1$ such that $\{p_1, q_l\}$ is an edge of G, we let $\nu(q_1) = 2$. Otherwise, we let $\nu(p_1) = 2$. We continue the same way: If the r-th vertex was already chosen, we consider the first p_i and the first q_j that are not yet assigned a number. If p_i is adjacent in G to some q_l, $l > j$, we let $\nu(q_j) = r + 1$. Otherwise, we let

$\nu(p_i) = r + 1$. When $\nu(p_i)$ is defined for all the vertices of the path P, there may still be some vertices on the path Q which were not yet assigned a number. We complete the numbering by keeping these vertices in their natural order on Q. For the NCC graph drawn above, the new order is:

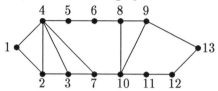

Given this order, we denote by G_i the subgraph induced by $\{i, \ldots, n\}$. Then G_i is either an NCC graph, or an NCC graph with one or two "tails" attached to it. In this subgraph, the degree of i is at most 2, and if i has degree 2 in G_i, then $i + 1$ is one of the two vertices adjacent to i. Given a diminishable DNN matrix realization of G, we may therefore try deleting successively vertices of degree 2 or 1. The process will work, due to the next two results:

Theorem 2.23 *If A is doubly nonnegative and $G(A)$ is an NCC graph, then A is diminishable if and only if A is singular.*

Proof. We already know that every diminishable matrix is singular, so we only have to prove the reverse implication. Suppose A is a doubly nonnegative singular matrix and $G(A)$ is an NCC graph whose vertices are ordered as above. The vertex 1 is adjacent to 2 and some $l \geq 3$. The matrix A therefore is in the form (2.19), where S has one nonzero entry in its first row. We will show that $A[1, \ldots, n - 2 | \ldots, n]$ is necessarily nonsingular, and since A is positive semidefinite, M is nonsingular. This, together with the singularity of A would imply that A is diminishable.

To prove that $A[1, \ldots, n - 2 | 3, \ldots, n]$ is nonsingular, we reorder the columns of A. Let $\sigma : \{1, \ldots, n - 2\} \to \{3, \ldots, n\}$ be defined as follows:

$$\begin{aligned}
\sigma(1) &= l \\
\sigma(i) &= \nu(p_{h+1}) \quad \text{if } i = \nu(p_h), h < m - 1 \\
\sigma(i) &= \nu(q_{r+1}) \quad \text{if } i = \nu(q_r), r < k - 1 \\
\sigma(i) &= n \quad\quad\ \ \text{if } i = \nu(p_{m-1})
\end{aligned}$$

Now let A^σ be the matrix obtained from A by leaving the first two columns as they are, and for $1 \leq i \leq n - 2$ letting column $i + 2$ of A^σ be column $\sigma(i)$ of A. We claim that $A^\sigma[1, \ldots, n - 2 | 3, \ldots, n]$ is a lower triangular matrix with all its diagonal elements nonzero. Indeed, for $1 \leq i \leq n - 2$

$$A^{\sigma}_{i,i+2} = a_{i,\sigma(i)} > 0,$$

since if i is on the path P, $\sigma(i)$ is the next vertex in the natural order of P, and if i is on Q, $\sigma(i)$ is the next vertex in the natural order of Q. If $j > i + 2$, then $\{i, \sigma(j)\}$ is not an edge of G. This is clear if both i and j are internal vertices of P, or both are internal vertices of Q, and it is clear also if $i = 1$. If $i = \nu(p_h)$ for some h and $j = \nu(q_r)$ for some r, then $\sigma(j) = \nu(q_{r+1})$, and by the definition of ν, p_h is not adjacent to q_{r+1}. If $i = \nu(q_r)$ for some r and $j = \nu(p_h)$ for some h, then by the definition of ν, some vertex p_s, $s \leq h$ is adjacent to some q_t, $t > r$. By (2.23), p_h cannot be adjacent to q_r.   □

Proposition 2.10   *If A is an $n \times n$ diminishable doubly nonnegative matrix in block form (2.19) and $d_{G(A)}(1) \leq 2$, then $A^{[1]}$ is singular.*

Proof.   Since $A = A^{[1/2]} + \alpha A^R$ and A^R is positive semidefinite,

$$\operatorname{rank} A^{[1/2]} \leq \operatorname{rank} A \leq n - 1.$$

By Theorem 1.20,

$$\operatorname{rank} A^{[1]} = \operatorname{rank} A^{[1/2]} - 1 \leq n - 2,$$

so $A^{[1]}$ is singular.   □

Given a singular doubly nonnegative matrix A whose graph is an NCC graph, we may define for $r \geq 2$,

$$A^{[r]} = (A^{[r-1]})^{[1]}.$$

By Theorems 2.22 and 2.23 and Proposition 2.10, the matrix $A^{[r-1]}$ is completely positive if and only if $A^{[r]}$ is completely positive. We therefore have the following result, which is an easy consequence of the previous discussion.

Corollary 2.12   *Let A be a singular $n \times n$ doubly nonnegative matrix whose graph is an NCC graph. Then A is completely positive if and only if for every $1 \leq r \leq n - 4$ the matrix $A^{[r]}$ is nonnegative.*

Exercises

2.32   Prove the statement in Example 2.13.

2.33   Prove Proposition 2.9, and use it to show that a matrix which is permutationally similar to the matrix A of Example 2.14 does not have property PLSS.

2.34   Use the fact that every $n \times n$ doubly nonnegative matrix, $n \leq 4$, is completely positive to show that every such matrix has property PLSS.

2.35   Prove (without using Theorem 2.15) that if a graph G is T_n and the vertex 1 is incident with the common base of the $n - 2$ triangles, then every DNN matrix realization of G has property PLSS.

2.36   Check which of the following doubly nonnegative matrices is completely positive:

(a)
$$\begin{pmatrix} 12 & 2 & 2 & 6 & 4 & 2 \\ 2 & 23 & 7 & 9 & 6 & 4 \\ 2 & 7 & 5 & 7 & 0 & 0 \\ 6 & 9 & 7 & 13 & 0 & 0 \\ 4 & 6 & 0 & 0 & 8 & 6 \\ 2 & 4 & 0 & 0 & 6 & 5 \end{pmatrix}$$

(b)
$$\begin{pmatrix} 5 & 0 & 1 & 1 & 1 & 1 \\ 0 & 13 & 1 & 3 & 3 & 5 \\ 1 & 1 & 2 & 3 & 0 & 0 \\ 1 & 3 & 3 & 5 & 0 & 0 \\ 1 & 3 & 0 & 0 & 2 & 4 \\ 1 & 5 & 0 & 0 & 4 & 10 \end{pmatrix}$$

(c)
$$\begin{pmatrix} 7 & 1 & 2 & 2 & 1 & 1 \\ 1 & 12 & 1 & 3 & 3 & 5 \\ 2 & 1 & 2 & 3 & 0 & 0 \\ 2 & 3 & 3 & 5 & 0 & 0 \\ 1 & 3 & 0 & 0 & 2 & 4 \\ 1 & 5 & 0 & 0 & 4 & 10 \end{pmatrix}$$

2.37   Prove Corollary 2.11.

2.38   Construct an example of a nonsingular critical CP matrix realization of C_5.

2.39   Show that every CP realization of an odd cycle is not i-minimizable for every i.

2.40   Let

$$A = \begin{pmatrix} 3 & 3 & 3 & 0 & 0 & 0 \\ 3 & 11 & 1 & 4 & 1 & 0 \\ 3 & 1 & 7 & 2 & 1 & 0 \\ 0 & 4 & 2 & 6 & 4 & 2 \\ 0 & 1 & 1 & 4 & 5 & 4 \\ 0 & 0 & 0 & 2 & 4 & 4 \end{pmatrix}.$$

Show that

$$A^{[1]} = \begin{pmatrix} 6 & 0 & 4 & 1 & 0 \\ 0 & 2 & 2 & 1 & 0 \\ 4 & 2 & 6 & 4 & 2 \\ 1 & 1 & 4 & 5 & 4 \\ 0 & 0 & 2 & 4 & 4 \end{pmatrix},$$

and conclude that A is completely positive.

2.41   Check if the matrix

$$A = \begin{pmatrix} 2 & 2 & 2 & 0 & 0 & 0 \\ 2 & 7 & 1 & 4 & 1 & 0 \\ 2 & 1 & 7 & 1 & 1 & 0 \\ 0 & 4 & 1 & 6 & 4 & 2 \\ 0 & 1 & 1 & 4 & 5 & 4 \\ 0 & 0 & 0 & 2 & 4 & 4 \end{pmatrix}.$$

is completely positive.

2.42   Check if the matrix

$$A = \begin{pmatrix} 2 & 1 & 2 & 0 & 0 & 0 & 0 \\ 1 & 2 & 3 & 1 & 0 & 0 & 0 \\ 2 & 3 & 7 & 3 & 2 & 0 & 0 \\ 0 & 1 & 3 & 4 & 1 & 1 & 0 \\ 0 & 0 & 2 & 1 & 5 & 1 & 2 \\ 0 & 0 & 0 & 1 & 1 & 3 & 2 \\ 0 & 0 & 0 & 0 & 2 & 2 & 2 \end{pmatrix}.$$

is completely positive.

2.43   Let a graph G be odd cycle C plus one chord, connecting two vertices u and v such that $d_C(u, v) = 2$. Suppose that the vertices of C are

$\{1,2\},\{2,3\},\ldots,\{n-1,n\}$ and $\{1,n\}$, and that $u=2$, $v=n$. Let A be a DNN matrix realization of G. Show that A is completely positive if and only if

$$\det A \geq -4 \det A[1,2|1,n] \prod_{i=2}^{n-1} a_{i\,i+1}$$

Notes. Property PLSS was defined in [Salce and Zanardo (1993)], and the relevant subsection is based on that paper. Book graphs are defined and discussed in [Barioli (1998a)]. $\widehat{\text{CP}}$ graphs are called there *excellent graphs*, and DNN matrix realizations of such graphs are called *excellent matrices*. Corollary 2.11 was proved, along with other related results, in [Drew, Johnson and Lam (2001)]. The proof given there is direct, and does not depend on Theorem 2.18. Theorem 2.20 was proved in [Barioli (1998a)], and also in [Xu (2002)], using a different approach and relying on a theorem proved in [Xu and Li (2000)]. The subsection on deletion of a vertex of degree 2 is based on [Barioli (2001a)], where a triple consisting of a vertex of degree 2 and the two vertices adjacent to it is called a *dog-ear*, and NCC graphs are called *non crossing cycles*. Exercises 2.36 and 2.40–2.42 are taken from [Barioli (1998a)] and [Barioli (1998b)]. Exercise 2.43 is based on [Xu (2002)]. Completely positive matrices of order 5 are studied in [Xu (2001)].

2.7 Square factorizations

By Remark 2.5, small (order ≤ 4) completely positive matrices have square factorizations. That is, such a matrix A can be represented as BB^T where B is a nonnegative square matrix. This is generally not true for larger completely positive matrices.

Example 2.17 The matrix

$$A = \begin{pmatrix} 2 & 0 & 0 & 1 & 1 \\ 0 & 2 & 0 & 1 & 1 \\ 0 & 0 & 2 & 1 & 1 \\ 1 & 1 & 1 & 2 & 0 \\ 1 & 1 & 1 & 0 & 2 \end{pmatrix}$$

is completely positive, but does not have a square nonnegative decomposition.

In this section we discuss conditions which guarantee that A does have a nonnegative square factorization BB^T. Most of the section deals with triangular factorizations.

Definition 2.11 A matrix A is a *UL-completely positive* matrix if there exists a nonnegative upper triangular matrix B such that $A = BB^T$. A is a *LU-completely positive matrix* if there exists a nonnegative lower triangular matrix B such that $A = BB^T$.

Example 2.18 By Example 2.3, every 2×2 doubly nonnegative matrix is *UL*-completely positive. By the very same argument, every 2×2 doubly nonnegative matrix is also *LU*-completely positive: If

$$A = \begin{pmatrix} a & b \\ b & c \end{pmatrix}$$

is doubly nonnegative, then

$$A = \begin{pmatrix} 0 & 0 \\ 0 & \sqrt{c} \end{pmatrix}^2$$

if $a = 0$, and

$$A = \begin{pmatrix} \sqrt{a} & 0 \\ b/\sqrt{a} & \sqrt{c - b^2/a} \end{pmatrix} \begin{pmatrix} \sqrt{a} & 0 \\ b/\sqrt{a} & \sqrt{c - b^2/a} \end{pmatrix}^T$$

if $a > 0$.

In this section we will see that a 3×3 doubly nonnegative matrix is either *UL*-completely positive or *LU*-completely positive, but not necessarily both. But for $n \geq 4$, an $n \times n$ completely positive matrix may be neither *UL*- nor *LU*- completely positive.

The following theorem plays a crucial role in determining whether a given completely positive matrix with positive diagonal entries is *UL*-completely positive.

Theorem 2.24 *Let A be an $n \times n$ doubly nonnegative matrix such that a_{nn} is positive. Then A is a UL-completely positive if and only if $A/[n]$ is UL-completely positive.*

Proof.   Let $A = UU^T$ where U is an upper triangular nonnegative matrix. Partition U as

$$U = \begin{pmatrix} K & \mathbf{u} \\ \mathbf{0}^T & u_{nn} \end{pmatrix},$$

where K is of order $n - 1$. Then $A/[n] = KK^T$.

Conversely, if $A/[n] = KK^T$, where K is a nonnegative upper triangular matrix, partition A as

$$A = \begin{pmatrix} A_{11} & \mathbf{a} \\ \mathbf{a}^T & a_{nn} \end{pmatrix},$$

where $\mathbf{a} \in \mathbf{R}_+^{n-1}$, and A_{11} is $(n-1) \times (n-1)$. Let

$$U = \begin{pmatrix} K & \mathbf{a}/\sqrt{a_{nn}} \\ \mathbf{0}^T & \sqrt{a_{nn}} \end{pmatrix},$$

then $A = UU^T$.   □

For a sufficient condition for UL-complete positivity we need the following terminology:

Definition 2.12   Let A be an $n \times n$ matrix. Let $i, j, i_2, ..., i_k$ be indices such that $1 \le i \ne j < i_2 < ... < i_k$ and let $\alpha = \{i_2, ..., i_k\}$. Then the $k \times k$ minor $\det A[\{i\} \cup \alpha | \{j\} \cup \alpha]$ is called a *left almost principal minor of* A.

Definition 2.13   Let A be an $n \times n$ matrix. Let $i, j, i_1, ...i_{k-1}$ be indices such that $i_1 < ... < i_{k-1} < i \ne j \le n$ and let $\alpha = \{i_1, ..., i_{k-1}\}$. Then the $k \times k$ minor $\det A[\alpha \cup \{i\} | \alpha \cup \{j\}]$ is called a *right almost principal minor* of A.

We can now state:

Theorem 2.25   *If A is an $n \times n$ doubly nonnegative matrix, $n \ge 3$, and all its left almost principal minors are nonnegative, then A is UL-completely positive.*

Proof.   The proof is by induction. Since any 2×2 doubly nonnegative matrix is UL-completely positive (Example 2.18), we may consider the claim to hold for $n = 2$.

Let $n \geq 3$ and suppose the theorem holds for $n - 1$. Let A be an $n \times n$ doubly nonnegative matrix with nonnegative left almost principal minors. Then

$$A = \begin{pmatrix} A_{11} & \mathbf{a} \\ \mathbf{a}^T & a_{nn} \end{pmatrix},$$

where $\mathbf{a} \in \mathbf{R}_+^{n-1}$. If $a_{nn} = 0$, then $\mathbf{a} = \mathbf{0}$, and the claim follows from the induction hypothesis. If $a_{nn} > 0$, it suffices to prove that $A/[n]$ is UL-completely positive. The matrix $A/[n]$ is doubly nonnegative — positive semidefinite by Theorem 1.19, and nonnegative since, by Proposition 1.13,

$$(A/[n])_{ij} = \frac{\det A[i, n | j, n]}{a_{nn}}.$$

If $A/[n]$ is at least 3×3, then all its left almost principal minors are nonnegative: For any $\alpha \subseteq \{1, \ldots, n - 1\}$,

$$(A/[n])[\{i\} \cup \alpha | \{j\} \cup \alpha] = A[\alpha \cup \{i, n\} | \alpha \cup \{j, n\}]/[n].$$

So by Theorem 1.17,

$$\det\left((A/[n])[\{i\} \cup \alpha | \{j\} \cup \alpha]\right) = \frac{\det A[\alpha \cup \{i, n\} | \alpha \cup \{j, n\}]}{a_{nn}} \geq 0$$

for every $1 \leq i \neq j < \min \alpha$. Hence $A/[n]$ is an $(n - 1) \times (n - 1)$ doubly nonnegative matrix with nonnegative left almost principal minors, and by the induction hypothesis it is UL-completely positive. $\qquad\Box$

We leave it as an exercise to show that the converse of Theorem 2.25 is not true (see Exercise 2.45).

Analogous results hold for LU-completely positive matrices. For example,

Theorem 2.26 *An $n \times n$ doubly nonnegative matrix, $n \geq 3$, whose right almost principal minors are all nonnegative is LU-completely positive.*

To prove the promised result concerning 3×3 doubly nonnegative matrices we need the following lemma:

Lemma 2.7 *If A is a 3×3 doubly nonnegative matrix, then at least two of the following three inequalities hold:*

$$a_{11}a_{23} \;\geq\; a_{12}a_{13}$$
$$a_{22}a_{13} \;\geq\; a_{12}a_{23}$$

$$a_{33}a_{12} \ \geq \ a_{13}a_{23}.$$

Proof. Suppose that two of these inequalities don't hold, say,

$$a_{11}a_{23} < a_{12}a_{13} \text{ and } a_{22}a_{13} < a_{12}a_{23}.$$

Then

$$a_{11}a_{23}a_{22}a_{13} < a_{12}a_{13}a_{12}a_{23}.$$

But this implies that $a_{11}a_{22} < a_{12}^2$, which is impossible in a positive semi-definite A. □

Corollary 2.13   *If A is a 3×3 doubly nonnegative matrix, then A is UL-completely positive or LU-completely positive.*

Proof.   If any diagonal entry of A is zero, the result is true by Example 2.18. So suppose $a_{ii} > 0$ for $i = 1, 2, 3$. The only left almost principal minors of A are

$$\det A[1, 3|2, 3] = \det A[2, 3|1, 3] = a_{33}a_{12} - a_{13}a_{23},$$

and the only right almost principal minors of A are

$$\det A[1, 3|1, 2] = \det A[1, 2|1, 3] = a_{11}a_{23} - a_{12}a_{13}.$$

By Lemma 2.7, either the left almost principal minors are nonnegative, or the right almost principal minors are, or both. The result therefore follows from Theorems 2.25 and 2.26. □

Example 2.19 A can be LU-completely positive and not UL-completely positive. For example, consider the matrix

$$A = \begin{pmatrix} 2 & 1 & 3 \\ 1 & 3 & 2 \\ 3 & 2 & 5 \end{pmatrix}.$$

It is LU-completely positive by Theorem 2.26, but it is not UL-completely positive since $A/[3]$ is not nonnegative.

Corollary 2.14   *Every symmetric totally nonnegative matrix is (LU- and UL-) completely positive.*

Corollary 2.15   *Symmetric nonnegative Cauchy matrices, and in particular Hilbert matrices, are completely positive.*

UL-complete positivity is related to the concept of inverse M-matrix (the inverse of an M-matrix).

Theorem 2.27 *A nonsingular doubly nonnegative matrix is an inverse M-matrix if and only if PAP^T is UL-completely positive for every permutation matrix P.*

Proof. If A^{-1} is a symmetric M-matrix, then there exists a lower triangular M-matrix L such that $A^{-1} = LL^T$ (see Exercise 1.29). Hence $A = UU^T$ where $U = (L^T)^{-1}$ is a nonnegative upper triangular matrix. If A^{-1} is an M-matrix and P is a permutation matrix, then $(PAP^T)^{-1} = PA^{-1}P^T$ is also an M-matrix, so this proves the "only if" part.

For the converse, we show that if $A = UU^T$, where U is an upper triangular nonnegative matrix, then the 21 entry of A^{-1} is nonpositive. This entry is equal to $-\det A(1|2)/\det A$. It is nonpositive since $\det A(1|2) = (\prod_{i=2}^n u_{ii})\, u_{12} (\prod_{i=3}^n u_{ii}) \geq 0$ and $\det A = (\prod_{i=1}^n u_{ii})^2$ is positive. So if PAP^T is UL-completely positive for every permutation matrix P, then all the off-diagonal entries of A^{-1} are nonpositive. $\qquad \square$

The class of matrices with nonnegative square factorizations includes, of course, matrices that are squares of symmetric nonnegative matrices. Square factorizations of this type have nothing to do with the triangular factorizations discussed so far. A matrix may be a square of a nonnegative symmetric matrix, but not be LU- or UL- completely positive, and vice versa. To conclude the section, we present one case in which it is possible to verify that A is a square of a symmetric nonnegative matrix.

Recall that a square nonnegative matrix A is *stochastic* if all its row sums are equal to 1. Such A is *doubly stochastic* if all its column sums are also equal to 1. That is, $A \geq 0$ is doubly stochastic if and only if

$$Ae = e \quad \text{and} \quad A^T e = e.$$

Theorem 2.28 *Let A be an $n \times n$ doubly nonnegative doubly stochastic matrix which has no diagonal entry greater than $1/(n-1)$. Then there exists a doubly nonnegative doubly stochastic matrix B such that $A = B^2$.*

Proof. Let B be the square root of A (see Remark 1.1). Since A is positive semidefinite and 1 is an eigenvalue of A with e as an associated eigenvector, and since B is positive semidefinite, 1 is also an eigenvalue of B and e is an eigenvector associated with it. So the sum of each row of B

(and each column, since B is symmetric) is 1. We need to show that B is nonnegative.

If $b_{i_0 j_0} < 0$, then the sum of the rest of the elements in the i_0-th row is greater than 1. Let

$$\mu = \sum_{\substack{j=1 \\ j \neq j_0}}^{n} b_{i_0 j} > 1.$$

By the Cauchy-Schwarz inequality,

$$\mu^2 = \left(\sum_{\substack{j=1 \\ j \neq j_0}}^{n} b_{i_0 j} \right)^2 \leq \left(\sum_{\substack{j=1 \\ j \neq j_0}}^{n} b_{i_0 j}^2 \right) (n-1).$$

This implies that

$$a_{i_0 i_0} = \sum_{j=1}^{n} b_{i_0 j}^2 \geq \sum_{\substack{j=1 \\ j \neq j_0}}^{n} b_{i_0 j}^2 \geq \frac{\mu^2}{n-1} > \frac{1}{n-1},$$

which contradicts the assumption on the diagonal elements of A.　　　　□

The requirement that the diagonal entries are not greater than $1/(n-1)$ cannot be dismissed — see Exercise 2.50.

Exercises

2.44　Prove the claim in Example 2.17.

2.45　Show that

$$A = \begin{pmatrix} 6 & 5 & 3 & 0 \\ 5 & 11 & 4 & 0 \\ 3 & 4 & 2 & 0 \\ 0 & 0 & 0 & 0 \end{pmatrix}$$

is *UL*-completely positive but has a negative left almost principal minor, or construct an example of your own to demonstrate that the converse of Theorem 2.25 is not true.

2.46　Find an example of a 4×4 doubly nonnegative matrix which is a square of a nonnegative matrix, and is neither *LU*- nor *UL*- completely positive.

2.47   Let $A = H(5)$, the 5×5 Hilbert matrix. Find a nonnegative matrix B such that $A = BB^T$.

2.48   Let A be an $n \times n$ *UL*-completely positive matrix, $n \geq 3$. Show that $\det A[\{i\} \cup \alpha | \{j\} \cup \alpha] \geq 0$ for every $1 \leq i \neq j < \min \alpha$, for all $\alpha = \{k, \ldots, n\}$, $k = 3, \ldots, n$.

2.49   Let A be an $n \times n$ nonsingular doubly nonnegative matrix, $n \geq 2$. Then A is *UL*-completely positive if and only if

$$\det A[i, k, \ldots, n | j, k, \ldots, n] \geq 0$$

for every $1 \leq i, j < k$ and $2 \leq k \leq n$.

2.50   Let A be the doubly stochastic doubly nonnegative matrix

$$\begin{pmatrix} \frac{3}{4} & 0 & \frac{1}{4} \\ 0 & \frac{3}{4} & \frac{1}{4} \\ \frac{1}{4} & \frac{1}{4} & \frac{1}{2} \end{pmatrix}.$$

(a) Show that there is no square nonnegative matrix B such that $A = B^2$ (in particular, there is no doubly stochastic B such that $A = B^2$).

(b) Show that there is no upper triangular nonnegative matrix U such that $A = UU^T$.

(c) Find a lower triangular nonnegative matrix L such that $A = LL^T$.

Notes.   Most of the section is based on [Markham (1971)]. Lemma 2.7 appears in [Maxfield and Minc (1962)] as part of the proof that 3×3 doubly nonnegative matrices are completely positive. The M-matrix results are taken from [Ando (1987)]. The M-matrix results are taken from [Ando (1987)] and [Jacobson (1974)]. See also [Lau and Markham (1978)]. Theorem 2.28 and the example in Exercise 2.50 are taken from [Marcus and Minc (1962)].

2.8   Functions of completely positive matrices

Given a completely positive matrix A and a function f, it is natural to ask when is $f(A)$ also completely positive. But first we have to define what we mean by $f(A)$. The ordinary definition of a function of a matrix is given through its Jordan form. In the case of real symmetric matrices the definition becomes:

Definition 2.14   Given an $n \times n$ real symmetric matrix A, let U be an orthogonal matrix and $D = \mathrm{diag}(d_1, d_2, \ldots, d_n)$ a diagonal matrix such that $U^T A U = D$. If $d_1, d_2, \ldots, d_n$ belong to the domain of a function f, then $f(A)$ is defined by $f(A) = U \mathrm{diag}(f(d_1), f(d_2), \ldots, f(d_n)) U^T$.

In the case that $f(x) = \sum_{i=0}^{n} a_i x^i$ is a polynomial, this definition agrees with the natural definition $f(A) = \sum_{i=0}^{n} a_i A^i$. And if $f(x)$ admits a power series expansion, $f(x) = \sum_{i=0}^{\infty} a_i x^i$, and $\lim_{m \to \infty} \sum_{i=0}^{\infty} a_i A^i$ exists, then $f(A)$ equals that limit, that is, $f(A) = \sum_{i=0}^{\infty} a_i A^i$.

We also consider Hadamard functions of A:

Definition 2.15   Let $f(x)$ be a function whose domain includes the entries of a matrix A. The *Hadamard function of a matrix* A, $f_H(A)$, is defined entrywise by $(f_H(A))_{ij} = f(a_{ij})$.

Example 2.20   Let

$$A = \begin{pmatrix} 1 & 2 \\ 3 & 4 \end{pmatrix},$$

and let $f(x) = x^2 - 5x - 2$. Then

$$f_H(A) = \begin{pmatrix} -6 & -8 \\ -8 & -6 \end{pmatrix}.$$

The ordinary $f(A)$ is in this case the zero matrix.

Note that we use the natural convention $A^{(0)} = J$.

Theorem 2.29   *Suppose $f(x)$ is defined on $[0, \gamma)$, and has a series expansion $f(x) = \sum_{k=0}^{\infty} a_k x^k$ for $0 \le x < \gamma$, where $a_k \ge 0$ for every k. Then for every completely positive matrix A with $a_{ij} < \gamma$, $i, j = 1, \ldots, n$, the matrix $f_H(A)$ is also completely positive.*

Proof. $f_H(A) = \sum_{k=0}^{\infty} a_k A^{(k)}$. For every $k \ge 1$ the matrix $A^{(k)}$ is completely positive by Corollary 2.3, and $A^{(0)} = J$ is also completely positive. For every m, $\sum_{k=0}^{m} a_k A^{(k)}$ is completely positive by Theorem 2.2 and, by the same theorem,

$$f_H(A) = \lim_{m \to \infty} \left(\sum_{k=0}^{m} a_k A^{(k)} \right)$$

is completely positive. $\qquad\square$

Corollary 2.16 *If A is completely positive, then so is $\exp_H(A)$.*

Proof. The function $\exp(x) = e^x$ satisfies the assumptions of the previous theorem for $\gamma = \infty$. $\qquad\square$

The last corollary can be used to obtain a stronger result:

Theorem 2.30 *If A is a real positive semidefinite matrix, then $\exp_H(A)$ is completely positive.*

Proof. Since A is positive semidefinite, it can be represented as sum of rank 1 positive semidefinite matrices, $\sum_{k=1}^{m} \mathbf{x}_k \mathbf{x}_k^T$, where every $\mathbf{x}_k$ is a (not necessarily nonnegative) vector in $\mathbf{R}^n$. Thus $\exp_H(A)$ is the Hadamard product of the Hadamard exponents $\exp_H(\mathbf{x}_k \mathbf{x}_k^T)$. By Corollary 2.2 it is therefore enough to prove that for every $\mathbf{x} \in \mathbf{R}^n$, $\exp_H(\mathbf{x}\mathbf{x}^T)$ is completely positive. The ij entry of $\exp_H(\mathbf{x}\mathbf{x}^T)$ is $e^{x_i x_j}$. Let $M = \max_{1 \leq i \leq n}(|x_i|)$. Then

$$e^{x_i x_j} = e^{-M^2 + (M+x_i)(M+x_j) - Mx_i - Mx_j}.$$

Thus

$$\exp_H(\mathbf{x}\mathbf{x}^T) = e^{-M^2} \exp_H(\mathbf{y}\mathbf{y}^T) \circ \mathbf{z}\mathbf{z}^T,$$

for the following $\mathbf{y}, \mathbf{z} \in \mathbf{R}^n$: $y_i = M + x_i, z_i = e^{-Mx_i}$. Since both $\mathbf{y}$ and $\mathbf{z}$ are nonnegative, $\mathbf{y}\mathbf{y}^T$ and $\mathbf{z}\mathbf{z}^T$ are completely positive. By Corollary 2.16, $\exp_H(\mathbf{y}\mathbf{y}^T)$ is completely positive, and since $e^{-M^2} > 0$, it follows from Corollary 2.2 that $\exp_H(\mathbf{x}\mathbf{x}^T)$ is completely positive. $\qquad\square$

In the next definition we use $A^{(p)}$ to denote the matrix whose ij entry is a_{ij}^p, where p is any positive number.

Definition 2.16 A matrix A is *infinitely divisible* if it is doubly nonnegative and the matrix $A^{(p)}$ is positive semidefinite for every real $p > 0$.

Theorem 2.31 *Infinitely divisible matrices are completely positive.*

Proof. The equality

$$a_{ij} = \lim_{p \to 0+} \exp\left(\frac{1}{p}(a_{ij}^p - 1)\right)$$

implies that

$$A = \lim_{p \to 0+} e^{-1/p} \exp_H \left(\frac{1}{p} A^{(p)} \right).$$

By Theorem 2.30 $\exp_H \left(\frac{1}{p} A^{(p)} \right)$ is completely positive for every $p > 0$, hence by Theorem 2.2 A is also completely positive. $\qquad \square$

The converse of Theorem 2.31 is not true.

Example 2.21 For every $\varepsilon > 0$,

$$A_\varepsilon = \begin{pmatrix} 1+\varepsilon & 1+2\varepsilon & 1 \\ 1+2\varepsilon & 1+4\varepsilon & 1 \\ 1 & 1 & 1 \end{pmatrix} = \begin{pmatrix} 1 & \sqrt{\varepsilon} \\ 1 & 2\sqrt{\varepsilon} \\ 1 & 0 \end{pmatrix} \begin{pmatrix} 1 & 1 & 1 \\ \sqrt{\varepsilon} & 2\sqrt{\varepsilon} & 0 \end{pmatrix}$$

is completely positive, but for small ε it is not infinitely divisible, since $\det A_\varepsilon^{(p)}$ is negative for $0 < p < 1$ and small ε.

Results similar to Theorem 2.29 and its corollaries hold for ordinary functions of completely positive matrices. Here we have to rely on the fact that natural powers of completely positive matrices are completely positive (Corollary 2.4), instead of the corresponding result for Hadamard powers.

The results are stated here, and left as exercises:

Theorem 2.32 *Suppose $f(x)$ has a power series expansion*

$$f(x) = \sum_{i=0}^{\infty} a_i x^i \, , \ 0 \le x \le \gamma,$$

in which $a_i \ge 0$ for every i. If A is completely positive and $\|A\| \le \gamma$, then $f(A)$ is completely positive.

Corollary 2.17 *If A is a nonnegative symmetric matrix, then $\exp(A)$ is completely positive.*

Exercises

2.51   If A is a real positive semidefinite matrix and $|a_{ij}| < 1$ for every i and j, then the matrix $(1/(1 - a_{ij}))_{i,j=1}^{n}$ is completely positive.
Hint: $1/(1 - x) = \exp(-\log(1 - x))$.

2.52 If A is nonnegative and symmetric, and if for every positive q the matrix $(a_{ij}/(q + a_{ij}))_{i,j=1}^n$ is positive semidefinite, then A is completely positive.

Hint: Use the identity

$$x^p = \frac{\sin(\pi p)}{\pi} \int_0^\infty \frac{x}{t + x} t^{p-1} dt,$$

which is valid for every $0 < p < 1$.

2.53 Prove Theorem 2.32.

2.54 Prove Corollary 2.17.

2.55 Show that the equivalent of Exercise 2.51 for functions of matrices does not hold. That is, give an example of a positive semidefinite matrix A such that $\|A\| < 1$ and $(I - A)^{-1}$ is not completely positive.

Show that if A is completely positive and $\|A\| < 1$, then $(I - A)^{-1}$ is completely positive.

Notes. This section is based on [Ando (1991)]. Infinitely divisible matrices were defined and characterized by R. A. Horn, see [Horn (1967)] and [Horn (1969)]. The name comes from the related concept of infinitely divisible function, see [Fitzgerald and Horn (1977)]. The integral identity in the hint for Exercise 2.52 can be derived by methods of complex analysis — see, *e.g.*, [Beardon (1979)]. It is also true that if A is a real positive semidefinite matrix, and if $A(pI + A)^{-1}$ is doubly nonnegative for every $p > 0$, then A is completely positive — see [Ando (1991)]).

2.9 The CP completion problem

The PSD completion problem was described in Section 1.7. Here we describe the analogous CP completion problem.

We will need the terms defined in Section 1.7, such as partial symmetric matrix, partial PSD matrix, and specification graph. Recall that in a partial symmetric matrix all the diagonal entries are specified. Here we define

Definition 2.17 A *partial CP matrix* is a partial symmetric matrix in which all the fully specified principal submatrices are completely positive. A *CP completion* of A is a completion of A which is completely positive.

If A is a partial CP matrix with specification graph G, we say that A is a *G-partial CP matrix*.

The question is: which partial CP matrices have a CP completion? Our first aim in this section is to characterize the graphs G such that every G-partial CP matrix is CP completable. By the following remark it suffices to characterize the *connected* graphs with this property.

Remark 2.12 A graph G has the property that all its partial CP matrix realizations can be completed to a completely positive matrix if and only if all its components have this property.

The main theorem in this section is:

Theorem 2.33 *Every partial CP matrix realization of a connected graph G can be completed to a completely positive matrix if and only if G is a block-clique graph.*

Recall that a block-clique graph is a graph in which each block is a clique (see Section 1.5). Before we prove the theorem, we consider two examples of partial CP matrices that cannot be completed to a doubly nonnegative matrix (let alone to a completely positive matrix).

Example 2.22 Let

$$
A = \begin{pmatrix}
1 & 1 & ? & & \cdots & ? & 0 \\
1 & 1 & 1 & ? & & ? & ? \\
? & 1 & 1 & \ddots & \ddots & & \vdots \\
& ? & \ddots & \ddots & \ddots & ? & \\
\vdots & & \ddots & \ddots & 1 & 1 & ? \\
? & ? & & ? & 1 & 1 & 1 \\
0 & ? & \cdots & \cdots & ? & 1 & 1
\end{pmatrix}.
$$

It is the $k \times k$ partial symmetric matrix of Example 1.35, where the diagonal, first superdiagonal and first subdiagonal entries are equal to 1, the $1k$ and the $k1$ entries are equal to 0, and all other entries are unspecified. Clearly, the fully specified principal submatrices are the diagonal entries and the 2×2 principal submatrices, each of which is either J_2 or I_2, so A is a partial CP matrix. It was shown in Example 1.35 that A cannot be completed to a positive semidefinite matrix.

Example 2.23 Let A be the partial symmetric matrix

$$A = \begin{pmatrix} 2 & 3 & 0 & ? \\ 3 & 6 & 3 & 0 \\ 0 & 3 & 6 & 3 \\ ? & 0 & 3 & 2 \end{pmatrix}.$$

All fully specified submatrices are positive semidefinite of order less than 4 so by Theorem 2.4 they are completely positive. Hence A is a partial CP matrix. However, when $a_{14} = a_{41}$ are specified, $\det A = -27(a_{14} + 1)^2$, so A cannot be completed to a doubly nonnegative matrix. Observe that the specification graph of A is T_4.

We can now prove Theorem 2.33

Proof of Theorem 2.33.   If G is not chordal, then it contains an induced cycle of length $k \geq 3$. The partial PSD matrix in the proof of Theorem 1.39 is actually partial CP, and since it cannot be completed to a positive semidefinite matrix, it cannot be completed completely positive one.

If G is chordal but not block-clique, then it contains an induced T_4. In this case we use Example 2.23 to specify the entries of the double triangle, set all remaining diagonal entries to be 1, and all remaining off-diagonal specified entries to be 0. By Example 2.23 the partial CP matrix obtained that way has no doubly nonnegative completion. Hence to be CP completable, the graph must be block-clique.

Suppose now that G is a block-clique graph, and let A be any partial CP matrix realization of G. We have to show that A can be completed to a completely positive matrix. If G is a clique, then any "partial" CP matrix realization of G is already a completely positive matrix. Suppose that G has just two blocks, and let A be a partial CP matrix realization of G. Since complete positivity is invariant under permutation similarity or positive diagonal congruence we may assume that

$$A = \begin{pmatrix} B & \mathbf{c} & ? \\ \mathbf{c}^T & 1 & \mathbf{d} \\ ? & \mathbf{d} & E \end{pmatrix},$$

where

$$A_1 = \begin{pmatrix} B & \mathbf{c} \\ \mathbf{c}^T & 1 \end{pmatrix} \text{ and } A_2 = \begin{pmatrix} 1 & \mathbf{d} \\ \mathbf{d} & E \end{pmatrix}$$

are completely positive, and each question mark stands for a block of unspecified entries. Suppose

$$A_1 = \sum_{i=1}^{m_1} \mathbf{f}_i \mathbf{f}_i^T \text{ and } A_2 = \sum_{j=1}^{m_2} \mathbf{g}_j \mathbf{g}_j^T,$$

$\mathbf{f}_i \in \mathbf{R}_+^p$, $\mathbf{g}_j \in \mathbf{R}_+^q$, and $p+q-1 = n$ is the order of A. Partition the vectors $\mathbf{f}_i$ and $\mathbf{g}_j$:

$$\mathbf{f}_i = \left(\begin{array}{c} \mathbf{v}_i \\ \varphi_i \end{array} \right), \ \mathbf{g}_j = \left(\begin{array}{c} \psi_j \\ \mathbf{w}_j \end{array} \right),$$

where $\mathbf{v}_i$ is obtained from $\mathbf{f}_i$ by deleting the last entry, and $\mathbf{w}_j$ is obtained from $\mathbf{g}_j$ by deleting the first entry.

Define $m_1 m_2$ vectors $\mathbf{h}_{ij}$ in $\mathbf{R}_+^n$:

$$\mathbf{h}_{ij} = \left(\begin{array}{c} \mathbf{v}_i \psi_j \\ \varphi_i \psi_j \\ \varphi_i \mathbf{w}_j \end{array} \right).$$

Then

$$\sum_{i,j} \mathbf{h}_{ij} \mathbf{h}_{ij}^T = \left(\begin{array}{ccc} B & \mathbf{c} & \mathbf{c}\mathbf{d}^T \\ \mathbf{c}^T & 1 & \mathbf{d} \\ \mathbf{d}\mathbf{c}^T & \mathbf{d} & E \end{array} \right),$$

so A has a CP completion. For the general case, where G consists of k blocks, we proceed by induction on k. We use the fact that in a graph with k blocks, there is a block which contains only one cut vertex (see Exercise 1.31), and apply a construction similar to the one above. $\square$

One may also consider a DNN completion problem.

Definition 2.18 A *partial DNN matrix* is a partial PSD matrix whose specified entries are nonnegative. A *G-partial DNN matrix* is a partial DNN matrix whose specification graph is G.

It is interesting that although a matrix (of order > 4) may be doubly nonnegative but not completely positive, the DNN completion problem has the same qualitative solution as the CP completion problem. The interested reader is invited to prove this fact (Exercise 2.58).

Since the existence of induced cycles indicates that a graph is not chordal, it is natural to ask what additional conditions are required on the specified entries of a partial CP matrix whose specification graph contains cycles, to assure that it has a CP completion.

Let A be a C_n-partial CP matrix and let $D = \text{diag}(a_{11}^{1/2}, \ldots, a_{11}^{1/2})$. Then

$$DAD = \begin{pmatrix} 1 & \cos\theta_2 & ? & \cdots & ? & \cos\theta_1 \\ \cos\theta_2 & 1 & \cos\theta_3 & \ddots & ? & ? \\ ? & \ddots & \ddots & \ddots & \ddots & \vdots \\ \vdots & \ddots & \ddots & \ddots & \ddots & \vdots \\ ? & ? & \ddots & \cos\theta_{n-1} & 1 & \cos\theta_n \\ \cos\theta_1 & ? & \cdots & ? & \cos\theta_n & 1 \end{pmatrix}$$

is also partial CP. We refer to DAD as the *normalization* of A.

Definition 2.19   A partial DNN (CP) matrix *meets the cycle conditions* if in the normalization of each of its partial submatrices that correspond to a cycle (of length n),

$$2 \max_{1 \le i \le n} \theta_i \le \sum_{i=1}^{n} \theta_i.$$

We conclude this section by bringing, without proof, the following generalization of Theorem 2.33.

Theorem 2.34   *Let G be a graph. Then the following are equivalent:*

(a) *Every G-partial CP matrix which meets the cycle conditions has a completely positive completion.*

(b) *Every G-partial DNN matrix which meets the cycle conditions has a doubly nonnegative completion.*

(c) *All the blocks of G are cliques or cycles.*

Exercises

2.56   Prove the claim in Remark 2.12.

2.57   The partial symmetric matrix

$$\begin{pmatrix} k & 3 & 0 & ? \\ 3 & 6 & 3 & 0 \\ 0 & 3 & 6 & 3 \\ ? & 0 & 3 & k \end{pmatrix}$$

is partial CP for $k \geq 2$. Show that it is CP-completable if and only if $k \geq 3$, and find a CP completion for it when $k \geq 3$.

2.58 Prove: A graph G has the property that every partial DNN matrix realization of G can be completed to a doubly nonnegative matrix if and only if G is a block-clique graph.

Notes.

Theorem 2.33 and the theorem in Exercise 2.58 are due to [Drew and Johnson (1998)]. Theorem 2.34 was proved in [Drew, Johnson, Kilner and McKay (2000)]. Block-clique graphs were used in [Johnson and Smith (1996)], in a study of completion problems for M-matrices and inverse M-matrices. In [Drew, Johnson, Kilner and McKay (2000)] they are called 1-chordal graphs.

Chapter 3

CP rank

3.1 Definition and basic results

Definition 3.1 Let A be an $n \times n$ completely positive matrix. The minimal k such that $A = BB^T$ for some nonnegative $n \times k$ matrix B is the *cp-rank* of A. The cp-rank of A is denoted by cp-rank A.

Geometrically, if A is a completely positive matrix of rank r, then $A = W^T W$ where W is an $r \times n$ matrix. That is, $A = \mathrm{Gram}(\mathbf{w}_1, \ldots, \mathbf{w}_n)$ where $\mathbf{w}_1, \ldots, \mathbf{w}_n \in \mathbf{R}^r$ are the columns of W. The cp-rank of A is the minimal dimension k such that $\mathbf{w}_1, \ldots, \mathbf{w}_n$ can be isometrically embedded into $\mathbf{R}_+^k$.

Another way to look at it: cp-rank A is the minimal number of summands in a rank 1 representation of A, $A = \sum_{i=1}^{k} \mathbf{b}_i \mathbf{b}_i^T$, $\mathbf{b}_i \geq 0$. A rank 1 representation with $k = $ cp-rank A summands will be called a *minimal* rank 1 representation.

Using this last approach, it is easy to see that

Proposition 3.1 *If A and B are $n \times n$ completely positive matrices, then*

$$\text{cp-rank}\,(A + B) \leq \text{cp-rank}\,A + \text{cp-rank}\,B. \tag{3.1}$$

Another simple observation follows from Definition 3.1:

Proposition 3.2 *For every completely positive matrix A,*

$$\text{cp-rank}\,A \geq \text{rank}\,A. \tag{3.2}$$

Equality may occur in (3.2):

Theorem 3.1   *If A is a completely positive matrix and* rank $A \leq 2$, *then*

$$\text{cp-rank } A = \text{rank } A.$$

Proof.   This follows from Example 2.2 and the proof of Theorem 2.1. □

Theorem 3.2   *Let A be an $n \times n$ completely positive matrix. If $n \leq 3$, then*

$$\text{cp-rank } A = \text{rank } A.$$

Proof.   For $n \leq 2$ and for singular 3×3 completely positive matrices this follows from the previous theorem. So assume A is a 3×3 nonsingular completely positive matrix. Let δ_0 be the maximal $\delta > 0$ for which $A - \delta E_{33}$ is positive semidefinite ($\delta_0 = \det A / \det A(3)$). Then

$$A = \delta_0 E_{33} + (A - \delta_0 E_{33}),$$

and $A - \delta_0 E_{33}$ is a singular 3×3 doubly nonnegative matrix. Hence

$$\text{cp-rank } (A - \delta_0 E_{33}) = \text{rank } (A - \delta_0 E_{33}). \tag{3.3}$$

Combining (3.1), (3.2) and (3.3), we get that

$$\text{rank } A \leq \text{cp-rank } A \leq 1 + \text{cp-rank } (A - \delta_0 E_{33}) = 1 + \text{rank } (A - \delta_0 E_{33}) \leq 1 + 2,$$

which implies that cp-rank $A = \text{rank } A = 3$. □

For $n \times n$ matrices of rank r, where $r \geq 3$ and $n \geq 4$, the cp-rank may be greater than the rank.

Example 3.1   The matrix

$$\begin{pmatrix} 6 & 3 & 3 & 0 \\ 3 & 5 & 1 & 3 \\ 3 & 1 & 5 & 3 \\ 0 & 3 & 3 & 6 \end{pmatrix}$$

satisfies rank $A = 3$, cp-rank $A = 4$. To see that cp-rank $A = 4$, note that if $A = \sum_{i=1}^{m} \mathbf{b}_i \mathbf{b}_i^T$ is a minimal rank 1 representation of A, then $1 \in \text{supp } \mathbf{b}_i$ for at least one i, say $1 \in \text{supp } \mathbf{b}_1$. If $1 \notin \text{supp } \mathbf{b}_i$ for every $i \neq 1$, then

$$\mathbf{b}_1 \mathbf{b}_1^T = \begin{pmatrix} 6 & 3 & 3 & 0 \\ 3 & \frac{3}{2} & \frac{3}{2} & 0 \\ 3 & \frac{3}{2} & \frac{3}{2} & 0 \\ 0 & 0 & 0 & 0 \end{pmatrix}.$$

This implies that $(A - \mathbf{b}_1\mathbf{b}_1^T)_{23} < 0$. But $A - \mathbf{b}_1\mathbf{b}_1^T = \sum_{i=2}^m \mathbf{b}_i\mathbf{b}_i^T$ is completely positive, a contradiction. Hence 1 must belong to two of the supports supp $\mathbf{b}_i$, $i = 1, \ldots, m$. By symmetry, 4 must also belong to two of the supports. Add the fact that 1 and 4 cannot belong to the same support, to conclude that we necessarily have $m \geq 4$. That is,

$$\text{cp-rank}\, A \geq 4 > \text{rank}\, A = 3.$$

The next theorem shows that we actually have here cp-rank $A = 4$.

Theorem 3.3 *Let A be an $n \times n$ completely positive matrix, $n \leq 4$. Then cp-rank $A \leq n$.*

A few simple observations will help in the proof (and in future proofs as well). We state them here and leave the proofs as exercises.

Proposition 3.3 *If S is a nonsingular matrix with a nonnegative inverse, and SAS^T is completely positive, then A is also completely positive, and cp-rank $A \leq$ cp-rank (SAS^T).*

Corollary 3.1 *Let A be a completely positive matrix, D a positive diagonal matrix, and P a permutation matrix. Then*

$$\text{cp-rank}\, A = \text{cp-rank}\, DAD = \text{cp-rank}\, PAP^T.$$

We also use the following lemma:

Lemma 3.1 *If $x, y, z \geq 0$ and $0 \leq x + y - z \leq 1$, then there exist $0 \leq a, b \leq 1$ such that*

$$\sqrt{ab} \leq x, \quad \sqrt{(1-a)(1-b)} \leq y,$$

and

$$\sqrt{ab} + \sqrt{(1-a)(1-b)} = x + y - z.$$

Proof. Since $0 \leq x + y - z \leq 1$,

$$x + y - z = \cos\theta$$

for some $0 \leq \theta \leq \pi/2$. For every $0 \leq t \leq \pi/2 - \theta$,

$$\cos t \cos(t + \theta) + \sin t \sin(t + \theta) = \cos\theta$$

We show that there exists $0 \leq t \leq \pi/2 - \theta$ such that

$$\cos t \cos(t + \theta) \leq x, \tag{3.4}$$

and

$$\sin t \sin(t + \theta) \leq y. \tag{3.5}$$

If (3.5) holds for $t = \pi/2 - \theta$, then we are done, since (3.4) certainly holds for this value of t. Hence we consider the case that (3.5), which holds for $t = 0$, does not hold for $t = \pi/2 - \theta$. In that case, let t_0 be the maximal $0 \leq t \leq \pi/2 - \theta$ for which (3.5) holds. Then

$$\sin t_0 \sin(t_0 + \theta) = y.$$

Since

$$\cos t_0 \cos(t_0 + \theta) + \sin t_0 \sin(t_0 + \theta) = \cos \theta \leq x + y,$$

(3.4) also holds for t_0. Take $a = \cos^2 t_0$ and $b = \sin^2 t_0$. □

Proof of Theorem 3.3. For $n \leq 3$, the theorem follows from Theorem 3.2. (It also follows from the proof of Theorem 2.4 in Section 2.3, or Exercise 2.13, or Corollary 2.13.)

Let A is a 4×4 completely positive matrix. As in the proof of Theorem 2.4, we generate a doubly nonnegative matrix

$$C = \begin{pmatrix} 1 & c_{12} & c_{13} & c_{14} \\ c_{12} & 1 & c_{23} & c_{24} \\ c_{13} & c_{23} & 1 & 0 \\ c_{14} & c_{24} & 0 & 1 \end{pmatrix}.$$

That is, $C = D_2 S D_1 A D_1 S^T D_2$, where D_1 and D_2 are positive diagonal matrices, and S has a nonnegative inverse. In view of Proposition 3.3 and Corollary 3.1, it suffices to prove that cp-rank $C \leq 4$. As in the proof of Theorem 2.4, split C into $C = EE^T + F$, where

$$E = \begin{pmatrix} c_{13} & c_{14} \\ c_{23} & c_{24} \\ 1 & 0 \\ 0 & 1 \end{pmatrix}.$$

Then

$$F = \begin{pmatrix} f_{11} & f_{12} \\ f_{12} & f_{22} \end{pmatrix} \oplus 0_2.$$

Clearly EE^T is completely positive with cp-rank $= 2$, and

$$\begin{pmatrix} f_{11} & f_{12} \\ f_{12} & f_{22} \end{pmatrix} = C/C[3,4]$$

is positive semidefinite. Hence

$$f_{11}f_{22} - f_{12}^2 \geq 0. \tag{3.6}$$

If $f_{12} \geq 0$, then $F[1,2]$ is a 2×2 completely positive matrix. Hence cp-rank $F =$ cp-rank $F[1,2]$ is at most 2, and

$$\text{cp-rank } C \leq \text{cp-rank } EE^T + \text{cp-rank } F \leq 2 + 2 = 4.$$

So consider the case that $f_{12} < 0$. We write $EE^T = G_1 + G_2$, where

$$G_1 = \begin{pmatrix} c_{13} \\ c_{23} \\ 1 \\ 0 \end{pmatrix} \begin{pmatrix} c_{13} \\ c_{23} \\ 1 \\ 0 \end{pmatrix}^T \quad \text{and} \quad G_2 = \begin{pmatrix} c_{14} \\ c_{24} \\ 0 \\ 1 \end{pmatrix} \begin{pmatrix} c_{14} \\ c_{24} \\ 0 \\ 1 \end{pmatrix}^T.$$

G_1 and G_2 are rank 1 doubly nonnegative matrices, and

$$G_1 + G_2 + F = C.$$

In particular,

$$c_{13}c_{23} + c_{14}c_{24} + f_{12} = c_{12}. \tag{3.7}$$

By (3.6) and (3.7) we may apply Lemma 3.1 to

$$x = c_{13}c_{23}/\sqrt{f_{11}f_{22}}, \quad y = c_{14}c_{24}/\sqrt{f_{11}f_{22}}, \quad \text{and} \quad z = c_{12}/\sqrt{f_{11}f_{22}},$$

to get a and b such that

$$\sqrt{ab} \leq c_{13}c_{23}/\sqrt{f_{11}f_{22}}, \quad \sqrt{(1-a)(1-b)} \leq c_{14}c_{24}/\sqrt{f_{11}f_{22}},$$

and

$$\sqrt{ab} + \sqrt{(1-a)(1-b)} = -f_{12}/\sqrt{f_{11}f_{22}}.$$

For these a and b, let

$$\mathbf{x} = \begin{pmatrix} \sqrt{af_{11}} \\ -\sqrt{bf_{22}} \\ 0 \\ 0 \end{pmatrix} \text{ and }, \mathbf{y} = \begin{pmatrix} \sqrt{(1-a)f_{11}} \\ -\sqrt{(1-b)f_{22}} \\ 0 \\ 0 \end{pmatrix}.$$

Then each of the matrices $G_1 + \mathbf{x}\mathbf{x}^T$ and $G_2 + \mathbf{y}\mathbf{y}^T$ is a doubly nonnegative matrix of rank ≤ 2 and hence completely positive of cp-rank ≤ 2. Since $C = (G_1 + \mathbf{x}\mathbf{x}^T) + (G_2 + \mathbf{y}\mathbf{y}^T)$, this implies that

$$\text{cp-rank } C \leq \text{cp-rank }(G_1 + \mathbf{x}\mathbf{x}^T) + \text{cp-rank }(G_2 + \mathbf{y}\mathbf{y}^T) \leq 4.$$ $\qquad \square$

For $n \geq 5$, there exist $n \times n$ completely positive matrices A such that cp-rank $A > n$.

Example 3.2   Let A be any 5×5 completely positive matrix whose graph is $K_{2,3}$. Let $A = \sum_{i=1}^{m} \mathbf{b}_i \mathbf{b}_i^T$ be any rank 1 representation of A. The support of each $\mathbf{b}_i$ is a clique in $G(A) = K_{2,3}$, and the corresponding complete graphs necessarily cover all the edges of $G(A)$. Since the maximal cliques in $K_{2,3}$ are of size 2, and there are 6 such cliques (edges), we get that cp-rank $A \geq 6$.

The argument in Example 3.2 suggests a general lower bound on the cp-rank of a matrix, given in terms of its graph:

Remark 3.1   If $A = \sum_{i=1}^{m} \mathbf{b}_i \mathbf{b}_i^T$ is any rank 1 representation of a completely positive matrix A, then the supports of the vectors $\mathbf{b}_i$ are cliques in $G(A)$, and the induced complete subgraphs cover all the edges of $G(A)$ (that is, every edge of $G(A)$ is an edge of one of these complete subgraphs). So, with each graph G we associate the *edge-clique cover number*, the minimal number of complete subgraphs of G needed to cover all of G's edges. We denote this number by $c(G)$. Then

$$\text{cp-rank } A \geq c(G(A)). \tag{3.8}$$

Strict inequality may occur — see Example 3.1.

It is useful to define also the cp-rank of a graph:

Definition 3.2   For any graph G,

cp-rank $G = \max\{$cp-rank $A \mid A$ is a CP matrix realization of $G\}$.

The inequality cp-rank $A \geq$ rank A implies that cp-rank $G \geq |V(G)|$, and by Remark 3.1 we also have cp-rank $G \geq c(G)$. It is quite easy to see that

$$\min\{\text{cp-rank } A \mid A \text{ is a CP matrix realization of } G\} = c(G).$$

We leave the proof as an exercise.

There is no known method of computing the cp-rank of a general completely positive matrix. The next three sections contain results on three related problems:

(A) For $r \geq 3$, what is the maximal possible cp-rank of a rank r completely positive matrix?

(B) For $n \geq 5$, what is the maximal possible cp-rank of an $n \times n$ completely positive matrix?

(C) For which completely positive matrices A the equality cp-rank $A =$ rank A holds?

To the list of observations which are useful in the study of these questions we should add one more: If $A = A_1 \oplus A_2$ were A_1 and A_2 are completely positive, then rank $A =$ rank $A_1 +$ rank A_2 and cp-rank $A =$ cp-rank $A_1 +$ cp-rank A_2. Hence we may restrict our attention to matrices with connected graphs.

Exercises

3.1 Prove Proposition 3.1.

3.2 Prove Proposition 3.2.

3.3 Prove Proposition 3.3.

3.4 Prove Corollary 3.1.

3.5 Suppose that A_n is completely positive for every n, and that

$$A = \lim_{n \to \infty} A_n.$$

Show that

$$\text{cp-rank } A \leq \liminf_{n \to \infty} \text{cp-rank } A_n.$$

3.6   Show that for every $1 \leq k \leq 4$ there exists a critical 4×4 completely positive matrix A such that $G(A) = K_4$ and cp-rank $A = k$.
Hint: For $k = 4$, let

$$A_\varepsilon = \begin{pmatrix} 2 & 0 & 1 & 1 \\ 0 & 2 & 1 & 1 \\ 1 & 1 & 2 & 0 \\ 1 & 1 & 0 & 2 \end{pmatrix} + \varepsilon J_4.$$

Show that cp-rank A_ε must be 4 for some $\varepsilon > 0$.

3.7   Let A is an $n \times n$ completely positive matrix, $\mathbf{b} \in \mathbf{R}_+^n$, and

$$A' = \begin{pmatrix} \mathbf{b}^T A \mathbf{b} & \mathbf{b}^T A \\ A\mathbf{b} & A \end{pmatrix}.$$

Show that cp-rank $A' =$ cp-rank A, and if A is critical, so is A'.

3.8   Show that for every graph G,

$$\min\{\text{cp-rank } A \mid A \text{ is a CP matrix realization of } G\} = c(G).$$

3.9   Prove that if H is a subgraph of G, then cp-rank $H \leq$ cp-rank G.

Notes.   The cp-rank of a completely positive matrix A is called the *factorization index* of A in some papers, and denoted by $\varphi(A)$. In other papers it is denoted by $\#(A)$. The fact that small matrices have square factorizations was proved by [Maxfield and Minc (1962)]. The proof we give here for 4×4 matrices is based on [Ando (1991)]. Exercise 3.6 is based on [Zhang and Li (2002)].

A notion related to the cp-rank is the *nonnegative rank* of a nonnegative matrix A. It is the smallest number of summands in a representation of A as a sum of rank 1 nonnegative matrices. For completely positive matrices, the nonnegative rank is a lower bound for the cp-rank. The two can be different, since $\min\{m, n\}$ is an upper bound on the nonnegative rank of an $m \times n$ matrix (see [Cohen and Rothblum (1993)]).

3.2   Completely positive matrices of a given rank

This section deals with problem (A) mentioned in Section 3.1: Find an upper bound on the cp-ranks of completely positive matrices of rank r. If

A is a rank r completely positive matrix, then $A = W^T W$ where W is an $r \times n$ matrix of rank r. Since A is completely positive, there exists a matrix Q such that $Q^T W \geq 0$ and $QQ^T = I_r$. The cp-rank of A is the minimal number of columns in such Q. Throughout this section we denote the columns of such W by $\mathbf{w}_1, \ldots, \mathbf{w}_n$, and the columns of such Q by $\mathbf{q}_1, \ldots, \mathbf{q}_k$. The cone generated by $\mathbf{w}_1, \ldots, \mathbf{w}_n$ in $\mathbf{R}^r$ is denoted by $\mathcal{C}_W$:

$$\mathcal{C}_W = \text{cone}\,(\mathbf{w}_1, \ldots, \mathbf{w}_n). \tag{3.9}$$

Since $A = \text{Gram}(\mathbf{w}_1, \ldots, \mathbf{w}_n)$ is completely positive and rank $W = r$, the polyhedral cone $\mathcal{C}_W$ is pointed and solid. The nonnegativity of $Q^T W$ means that

$$\mathbf{q}_1, \ldots, \mathbf{q}_k \in \mathcal{C}_W^* \tag{3.10}$$

and the equality $QQ^T = I_r$ translates to

$$\sum_{i=1}^{k} \mathbf{q}_i \mathbf{q}_i^T = I_r. \tag{3.11}$$

Hence the cp-rank of A is the minimal number k of vectors in $\mathcal{C}_W^*$ which satisfy (3.11). We use this approach and these notations in all the proofs in this section.

Theorem 3.4 *Let A be a rank r completely positive matrix, $r \geq 2$, and let N be the maximal number of zeroes above the diagonal in a nonsingular principal submatrix of A. Then*

$$\text{cp-rank}\,A \leq \frac{r(r+1)}{2} - N.$$

Proof. Let A be a rank r $n \times n$ completely positive matrix. There exist an $r \times n$ matrix W such that $A = W^T W$ and an $r \times k$ matrix Q whose columns $\mathbf{q}_1, \ldots, \mathbf{q}_k$ satisfy (3.10) and (3.11). Let C be the cone generated in $\mathcal{S}_r$ by $\{\mathbf{q}_i \mathbf{q}_i^T \mid i = 1, \ldots, k\}$, and

$$V = \text{Span}\,\{\mathbf{q}_i \mathbf{q}_i^T \mid i = 1, \ldots, k\} = \text{Span}\,(C) \subseteq \mathcal{S}_r.$$

Let $d = \dim V$. By (3.11) $I_r \in C$, and since $\dim(\text{Span}\,(C)) = d$, Theorem 1.34 (Carathéodory's Theorem) implies that I_r can be represented as a nonnegative combination of at most d of the $\mathbf{q}_i \mathbf{q}_i^T$'s. Without loss of generality,

$$I_r = \sum_{i=1}^{d} \alpha_i \mathbf{q}_i \mathbf{q}_i^T \quad , \ \alpha_i \geq 0, \ i = 1, \dots, d.$$

But then the matrix $P = (\sqrt{\alpha_1}\mathbf{q}_1 \dots \sqrt{\alpha_d}\mathbf{q}_d)$ is an $r \times d$ matrix such that $PP^T = I_r$ and $P^T W \geq 0$. Hence cp-rank $A \leq d$. To complete the proof we show that $d \leq \frac{r(r+1)}{2} - N$.

Let $A[\alpha]$, $\alpha \subseteq \{1, \dots, n\}$, be a nonsingular principal submatrix of A with N zeroes above the diagonal. Without loss of generality assume that $\alpha = \{1, \dots, h\}$. There are N pairs (p, q), $p, q \in \alpha$ and $p < q$, for which $a_{pq} = 0$, i.e., $\mathbf{w}_p^T \mathbf{w}_q = 0$. For each of these N pairs define a linear functional $f_{p,q} : \mathcal{S}_r \to \mathbf{R}$,

$$f_{p,q}(X) = \mathbf{w}_p^T X \mathbf{w}_q.$$

Obviously, $I_r \in \ker(f_{p,q})$ for each such pair. By (3.11)

$$\sum_{i=1}^{k} \mathbf{w}_p^T \mathbf{q}_i \mathbf{q}_i^T \mathbf{w}_q = \mathbf{w}_p^T I_r \mathbf{w}_q = 0, \tag{3.12}$$

and since $\mathbf{w}_p^T \mathbf{q}_i \mathbf{q}_i^T \mathbf{w}_q \geq 0$ for every i, (3.12) implies that $\mathbf{w}_p^T \mathbf{q}_i \mathbf{q}_i^T \mathbf{w}_q = 0$ for each i. That is,

$$\mathbf{q}_i \mathbf{q}_i^T \in \ker(f_{p,q}) \quad , i = 1, \dots, k.$$

Hence

$$V = \text{Span} \{\mathbf{q}_i \mathbf{q}_i^T \mid i = 1, \dots, k\} \subseteq \ker(f_{p,q})$$

for each such pair (p, q). It follows that $V \subseteq K$, where K is the intersection of all these kernels.

Let U be an $r \times r$ nonsingular matrix satisfying $U\mathbf{e}_i = \mathbf{w}_i$, $, i = 1, \dots, h$. Then $X \in K$ if and only if $\mathbf{e}_p^T U^T X U \mathbf{e}_q$ is zero for each of the N pairs (p, q). That is, $(U^T X U)_{pq} = 0$ for each such pair. Since $X \mapsto U^T X U$ is an isomorphism of $\mathcal{S}_r$ onto itself, this means that K is isomorphic to a subspace of $\mathcal{S}_r$ consisting of all $r \times r$ symmetric matrices with zeroes in N specified entries above the diagonal. In particular, $\dim K = \frac{r(r+1)}{2} - N$. And since $V \subseteq K$,

$$d = \dim V \leq \dim K = \frac{r(r+1)}{2} - N. \qquad \square$$

The next theorem provides a slight improvement of the bound in Theorem 3.4 in the case that A is a *positive* completely positive matrix. For the proof, we need one more observation.

Remark 3.2 If A is a rank r completely positive matrix, we may assume that $A = W^T W$ where W is $r \times n$ and every nonzero vector in $\mathcal{C}_W^*$ has a positive last entry. To see that, suppose $A = WW^T$, $W \in \mathbf{R}^{r \times n}$. Let α_i, $i = 1, \ldots, n$, be positive real numbers for which $\mathbf{w}_0 = \sum_{i=1}^n \alpha_i \mathbf{w}_i$ is a nonzero unit vector in $\mathcal{C}_W$, and let U be an orthogonal $r \times r$ matrix satisfying $U\mathbf{w}_0 = \mathbf{e}_r$. Then $A = (UW)^T(UW)$. If $0 \neq \mathbf{q} \in \mathcal{C}_{UW}^*$, then $\mathbf{q}_r = \mathbf{q}^T \mathbf{e}_r = \mathbf{q}^T U \mathbf{w}_0 = \sum_{i=1}^n \alpha_i \mathbf{q}^T U \mathbf{w}_i$. For every $i = 1, \ldots, n$, $\mathbf{q}^T U \mathbf{w}_i \geq 0$, and since $U\mathbf{w}_1, \ldots, U\mathbf{w}_n$ span $\mathbf{R}^r$ and $\mathbf{q} \neq 0$, at least one of the products $\mathbf{q}^T U \mathbf{w}_i$ is positive. Thus $\mathbf{q}_r > 0$.

Theorem 3.5 *For every rank r completely positive matrix A, $r \geq 2$,*

$$\text{cp-rank } A \leq \frac{r(r+1)}{2} - 1.$$

Proof. Suppose that A is an $n \times n$ rank r completely positive matrix. Let W be an $r \times n$ matrix such that $A = W^T W$, and let $\mathbf{q}_1, \ldots, \mathbf{q}_k \in \mathcal{C}_W^*$ satisfy (3.11). We may assume that the last component of each $\mathbf{q}_i$ is positive (by Remark 3.2), that $k \leq r(r+1)/2$ (by Theorem 3.4), and that the k rank 1 matrices $\mathbf{q}_1 \mathbf{q}_1^T, \ldots, \mathbf{q}_k \mathbf{q}_k^T$ are linearly independent (by Remark 1.6). If $k < r(r+1)/2$, there is nothing to prove, since cp-rank $A \leq k$. We therefore consider only the case $k = r(r+1)/2$.

For $i = 1, \ldots, k$, let $\mu_i = (\mathbf{q}_i)_r > 0$ and $\mathbf{p}_i = 1/\mu_i \mathbf{q}_i$. Then $(\mathbf{p}_i)_r = 1$ for each i, $\mathbf{p}_1 \mathbf{p}_1^T, \ldots, \mathbf{p}_k \mathbf{p}_k^T$ are linearly independent, and

$$\sum_{i=1}^k \mu_i^2 \mathbf{p}_i \mathbf{p}_i^T = I_r.$$

Define $\mathbf{p}_1(\beta) = (1 - \beta)\mathbf{p}_1 + \beta \mathbf{p}_2$, and consider the matrix equation

$$x_1 \mathbf{p}_1(\beta) \mathbf{p}_1^T(\beta) + \sum_{i=2}^k x_i \mathbf{p}_i \mathbf{p}_i^T = I_r. \tag{3.13}$$

Equation (3.13) translates into a system of k linear equations in the k unknowns $x_1, \ldots, x_k$. For $\beta = 0$ there is a unique solution $x_i = \mu_i^2$, $i = 1, \ldots, k$, and therefore the coefficients matrix of the system is nonsingular. For $\beta = 1$, the first two columns of the coefficients matrix are identical, so the matrix is singular. Let $0 < \beta_0 \leq 1$ be the smallest positive β for which the coefficients matrix is singular. The determinant $\Delta(\beta)$ of the coefficients matrix is continuous, hence for every $0 \leq \beta < \beta_0$ the coefficients matrix is nonsingular and its determinant $\Delta(\beta)$ has the same sign as $\Delta(0)$. For each such β the system has a unique solution, which may be represented by Cramer's rule:

$$\lambda_i(\beta) = \frac{\Delta_i(\beta)}{\Delta(\beta)} \quad, i = 1, \dots, k.$$

$\Delta_1(\beta)$ does not depend on β. The sign of $\lambda_1(\beta)$ is therefore fixed for every $0 \leq \beta < \beta_0$, i.e., $\lambda_1(\beta) > 0$. Since $\lim_{\beta \to \beta_0-} \Delta(\beta) = \Delta(\beta_0) = 0$, this implies

$$\lim_{\beta \to \beta_0-} \lambda_1(\beta) = \lim_{\beta \to \beta_0-} \frac{\Delta_1(\beta)}{\Delta(\beta)} = \infty.$$

Thus there exists $0 < \beta_1 < \beta_0$ for which $\lambda_1(\beta_1) > 1$. In particular

$$\lambda_1(\beta_1)(\mathbf{p}_1(\beta_1)\mathbf{p}_1^T(\beta_1))_{rr} + \sum_{i=2}^{k} \lambda_i(\beta_1)(\mathbf{p}_i\mathbf{p}_i^T)_{rr} = 1.$$

But $(\mathbf{p}_1(\beta_1))_r = (\mathbf{p}_2)_r = \dots = (\mathbf{p}_k)_r = 1$, hence

$$\sum_{i=1}^{k} \lambda_i(\beta_1) = 1.$$

Since $\lambda_1(\beta_1) > 1$, $\lambda_i(\beta_1) < 0$ for at least one of the other i's. Let β_2 be the largest $0 \leq \beta < \beta_1$ such that $\lambda_i(\beta) \geq 0$ for every $i = 1, \dots, k$. By the continuity of all the $\lambda_i(\beta)$'s, we have $\lambda_i(\beta_2) = 0$ for at least one i. So

$$\lambda_1(\beta_2)\mathbf{p}_1(\beta_2)\mathbf{p}_1^T(\beta_2) + \sum_{i=2}^{k} \lambda_i(\beta_2)\mathbf{p}_i\mathbf{p}_i^T = I_r$$

is a representation of I_r as a sum of less than $k = r(r+1)/2$ rank 1 matrices. Thus

$$\text{cp-rank } A \leq \frac{r(r+1)}{2} - 1.$$

$\square$

The next theorem shows that the upper bound in Theorem 3.5 is tight. The proof uses the following lemma.

Lemma 3.2 *Suppose there exist vectors* $\mathbf{p}_1, \ldots, \mathbf{p}_k \in \mathbf{R}^r$ *and positive real numbers* $\mu_1, \ldots, \mu_k$ *such that the following three conditions are satisfied:*

(a) *For each* $1 \leq i \leq k$, $\mathbf{p}_i^T = (\mathbf{u}_i^T, 1)$, *where each* $\mathbf{u}_i \in \mathbf{R}^{r-1}$ *satisfies* $\|\mathbf{u}_i\| = \sqrt{r-1}$.
(b) $\{\mathbf{p}_1\mathbf{p}_1^T, \ldots, \mathbf{p}_k\mathbf{p}_k^T \mid i = 1, \ldots, k\}$ *is a linearly independent set of matrices in* $\mathcal{S}_r$.
(c) $\sum_{i=1}^k \mu_i \mathbf{p}_i \mathbf{p}_i^T = I_r$.

Then there exists a completely positive matrix A *such that* rank $A = r$ *and* cp-rank $A = k$.

Proof. Let C be the convex cone generated in $\mathbf{R}^r$ by $\mathbf{p}_1, \ldots, \mathbf{p}_k$. Then

$$x_r \geq 0 \text{ for every } \mathbf{x} \in C, \text{ and } C \cap \{\mathbf{x} \in \mathbf{R}^r \mid x_r = 0\} = \{0\}. \qquad (3.14)$$

To see that the second part of (3.14) holds, note that if

$$\sum_{i=1}^k \alpha_i \mathbf{p}_i = \begin{pmatrix} \mathbf{y} \\ 0 \end{pmatrix}, \quad \mathbf{y} \in \mathbf{R}^{r-1}, \ \alpha_1, \ldots, \alpha_k \geq 0,$$

then equating the r-th component on both sides yields $\sum_{i=1}^k \alpha_i = 0$, and since the α_i's are all nonnegative, this implies that $\alpha_1 = \ldots = \alpha_k = 0$, hence $\mathbf{y} = 0$.

Another property of C is this:

$$\text{If } \begin{pmatrix} \mathbf{y} \\ 1 \end{pmatrix} \in C, \ \mathbf{y} \in \mathbf{R}^{r-1}, \text{ then } \|\mathbf{y}\| \leq \sqrt{r-1}, \qquad (3.15)$$

and equality occurs iff $\mathbf{y} = \mathbf{u}_i$ for some $i = 1, \ldots, k$.

To see that (3.15) holds, assume

$$\begin{pmatrix} \mathbf{y} \\ 1 \end{pmatrix} = \sum_{i=1}^k \alpha_i \mathbf{p}_i, \quad \alpha_1, \ldots, \alpha_k \geq 0.$$

Then $\sum_{i=1}^k \alpha_i = 1$ and $\sum_{i=1}^k \alpha_i \mathbf{u}_i = \mathbf{y}$, and therefore

$$\|\mathbf{y}\| \leq \sum_{i=1}^k \alpha_i \|\mathbf{u}_i\| = \sum_{i=1}^k \alpha_i \sqrt{r-1} = \sqrt{r-1}.$$

Since the $\mathbf{u}_i$'s are k different vectors with equal norm, equality occurs if and only if one of the α_i's is 1 and the others are all zero.

C is a polyhedral cone in $\mathbf{R}^r$ and (c) implies that it is reproducing. Hence C^* is also a reproducing polyhedral cone in $\mathbf{R}^r$. Let W be an $r \times n$ matrix whose columns generate C^*, and $A = W^T W$. Let Q be the matrix whose columns are $\sqrt{\mu_1}\mathbf{p}_1, \ldots, \sqrt{\mu_k}\mathbf{p}_k$. Then $Q^T W$ is $k \times n$, $Q^T W \geq 0$, and since $QQ^T = \sum_{i=1}^k \mu_i \mathbf{p}_i \mathbf{p}_i^T = I_r$,

$$A = W^T W = (Q^T W)^T (Q^T W).$$

That is, A is a completely positive matrix, $\operatorname{rank} A = r$ and cp-rank $A \leq k$.

To see that cp-rank $A = k$, suppose $\hat{Q}$ is an $r \times t$ matrix satisfying $\hat{Q}^T W \geq 0$ and $\hat{Q}\hat{Q}^T = I_r$, where $t = $ cp-rank A. Let $\hat{\mathbf{q}}_1, \ldots, \hat{\mathbf{q}}_t$ be $\hat{Q}$'s columns (necessarily, all nonzero). Let $\lambda_i = (\hat{\mathbf{q}}_i)_r$, $i = 1, \ldots, t$. By (3.14), $\lambda_i > 0$ for every i. Thus $\hat{\mathbf{q}}_i = \lambda_i \mathbf{b}_i$, where $\mathbf{b}_i^T = (\mathbf{v}_i^T, 1) \in C$. The equality

$$\sum_{i=1}^t \hat{\mathbf{q}}_i \hat{\mathbf{q}}_i^T = \sum_{i=1}^t \lambda_i^2 \mathbf{b}_i \mathbf{b}_i^T = I_r$$

implies that $\sum_{i=1}^t \lambda_i^2 = 1$, and since

$$\operatorname{trace}(I_r) = \sum_{i=1}^t \lambda_i^2 \operatorname{trace}(\mathbf{b}_i \mathbf{b}_i^T),$$

we have

$$r = \sum_{i=1}^t \lambda_i^2 \|\mathbf{v}_i\|^2 + \sum_{i=1}^t \lambda_i^2 = \sum_{i=1}^t \lambda_i^2 \|\mathbf{v}_i\|^2 + 1. \qquad (3.16)$$

By (3.15), $\|\mathbf{v}_i\|^2 \leq r - 1$ for each i, and thus from (3.16) it follows that $\|\mathbf{v}_i\|^2 = r - 1$ for each i. This in turn implies (again by (3.15)) that each $\mathbf{v}_i$ is one of $\mathbf{u}_1, \ldots, \mathbf{u}_k$. Hence

$$\{\mathbf{b}_1, \ldots, \mathbf{b}_t\} \subseteq \{\mathbf{p}_1, \ldots, \mathbf{p}_k\}.$$

Since $\{\mathbf{p}_1 \mathbf{p}_1^T, \ldots, \mathbf{p}_k \mathbf{p}_k^T \mid i = 1, \ldots, k\}$ is linearly independent, the two representations of I_r,

$$I_r = \sum_{i=1}^k \mu_i \mathbf{p}_i \mathbf{p}_i^T \quad, \quad I_r = \sum_{i=1}^t \lambda_i^2 \mathbf{b}_i \mathbf{b}_i^T$$

are the same. In particular, $t = k$. $\qquad\qquad\square$

We can now state and prove

Theorem 3.6 *For every $r \geq 2$ there exists a completely positive matrix A with rank $A = r$ and cp-rank $A = r(r+1)/2 - 1$.*

Proof. For $r = 2$ this follows from Theorem 3.1. We therefore consider the case $r \geq 3$. Let $k = r(r+1)/2 - 1$. We prove the theorem by constructing k vectors in $\mathbf{R}^r$ and k positive real numbers such that conditions (a),(b), and (c) of Lemma 3.2 are satisfied.

For every $2 \leq h \leq r$ and every $1 \leq l \leq h$ define $\mathbf{p}_{hl} \in \mathbf{R}^r$. For $l = 1$ let

$$
(\mathbf{p}_{h1})_i = \begin{cases} -ab^{h-2}, & i = 1 \\ -ab^{h-i}, & 2 \leq i \leq h - 1 \\ 0, & h \leq i \leq r - 1 \\ 1, & i = r \end{cases}
$$

and for $2 \leq l \leq h$ let

$$
(\mathbf{p}_{hl})_i = \begin{cases} 0, & i \leq l - 2 \\ ab^{h-l}, & i = l - 1 \\ -ab^{h-i}, & l \leq i \leq h - 1 \\ 0, & h \leq i \leq r - 1 \\ 1, & i = r \end{cases} ,
$$

Where $a = \sqrt{r - 1}$ and $b = 1/\sqrt{2}$.

We first show that these vectors satisfy (a) and (b). Clearly, each $\mathbf{p}_{hl}^T$ is of the form $(\mathbf{u}_{hl}^T, 1)$, and it is easy to check that $\|\mathbf{u}_{hl}\| = \sqrt{r - 1}$. That is, (a) is satisfied. We now show that $\{\mathbf{p}_{hl}\mathbf{p}_{hl}^T \mid h = 2, \ldots, r, l = 1, \ldots, h\}$ is a linearly independent set. First, we show that for each $2 \leq h \leq r$ the set $\{\mathbf{p}_{hl} \mid 1 \leq l \leq h\} \subseteq \mathbf{R}^r$ is linearly independent. Consider the matrix

$$
M_h = \begin{pmatrix} -ab^{h-2} & ab^{h-2} & 0 & 0 & \cdots & 0 & 0 \\ -ab^{h-2} & -ab^{h-2} & ab^{h-3} & 0 & \cdots & 0 & 0 \\ -ab^{h-3} & -ab^{h-3} & -ab^{h-3} & ab^{h-4} & \cdots & 0 & 0 \\ \vdots & \vdots & \vdots & \vdots & \ddots & \vdots & \vdots \\ -ab^2 & -ab^2 & -ab^2 & -ab^2 & \cdots & ab & 0 \\ -ab & -ab & -ab & -ab & \cdots & -ab & a \\ 1 & 1 & 1 & 1 & \cdots & 1 & 1 \end{pmatrix}.
$$

M_h consists of rows $1, \ldots, h-1, r$ of the matrix $(\mathbf{p}_{h1} \ldots \mathbf{p}_{hh})$. Divide each row by the absolute value of the first element in the row, to get

$$
\begin{pmatrix}
-1 & 1 & 0 & 0 & \cdots & 0 & 0 \\
-1 & -1 & 1/b & 0 & \cdots & 0 & 0 \\
-1 & -1 & -1 & 1/b & \cdots & 0 & 0 \\
\vdots & \vdots & \vdots & \vdots & \ddots & \vdots & \vdots \\
-1 & -1 & -1 & -1 & \cdots & 1/b & 0 \\
-1 & -1 & -1 & -1 & \cdots & -1 & 1/b \\
1 & 1 & 1 & 1 & \cdots & 1 & 1
\end{pmatrix}.
$$

The determinant of this $h \times h$ matrix is $-2(-1/b-1)^{h-2}$ (prove by induction on h). Hence

$$
\det(M_h) = -2a^{h-1}b^{\frac{h^2-h-2}{2}} \left(-1/b - 1\right)^{h-2} \neq 0
$$

We can now prove, by induction on m, that for each $2 \leq m \leq r$ the set $\{\mathbf{p}_{hl}\mathbf{p}_{hl}^T \mid h = 2, \ldots, m, l = 1, \ldots, h\}$ is linearly independent. For $m = 2$ this follows from the fact that $\{\mathbf{p}_{21}, \mathbf{p}_{22}\}$ are linearly independent. Let $2 < m \leq r$, and suppose $\{\mathbf{p}_{hl}\mathbf{p}_{hl}^T \mid h = 2, \ldots, m-1, l = 1, \ldots, h\}$ is linearly independent. Assume

$$
\sum_{h=2}^{m} \sum_{l=1}^{h} \alpha_{hl} \mathbf{p}_{hl} \mathbf{p}_{hl}^T = 0. \tag{3.17}
$$

For every $2 \leq h \leq m-1$ and $1 \leq l \leq h$

$$
\mathrm{supp}(\mathbf{p}_{hl}) \subseteq \{1, \ldots, m-2, r\}.
$$

Hence the $(m-1)$-th column of $\sum_{h=2}^{m-1} \sum_{l=1}^{h} \alpha_{hl} \mathbf{p}_{hl} \mathbf{p}_{hl}^T$ is zero. Equating the $(m-1)$-th columns on both sides of (3.17) we therefore get that

$$
\sum_{l=1}^{m} \alpha_{ml} (\mathbf{p}_{ml})_{m-1} \mathbf{p}_{ml} = 0.
$$

Since $\{\mathbf{p}_{m1}, \ldots, \mathbf{p}_{mm}\}$ is linearly independent and $(\mathbf{p}_{ml})_{m-1} \neq 0$ for every $l = 1, \ldots, m$, we get that $\alpha_{ml} = 0$ for every l. Hence (3.17) means

$$
\sum_{h=2}^{m-1} \sum_{l=1}^{h} \alpha_{hl} \mathbf{p}_{hl} \mathbf{p}_{hl}^T = 0,
$$

and by the induction hypothesis this implies that $\alpha_{hl} = 0$ for every $h = 1, \ldots, m-1, l = 1, \ldots, h$. This completes the proof that (b) is satisfied.

Finally, we define μ_{hl}, $h = 2, \ldots, r, l = 1, \ldots, h$, and show that (c) is satisfied. The numbers μ_{hl} are defined in the following table, where in addition to a and b we have $p = 2 - \sqrt{2}$.

	$l = 1, 2$	$l = 3, \ldots, h$
$h = 2$	$\mu_{hl} = \frac{b^2 p}{a^2}$	—
$h = 3, \ldots, r-1$	$\mu_{hl} = \frac{b^2 p^{h-2}}{a^2}$	$\mu_{hl} = \frac{b p^{h-l+1}}{a^2}$
$h = r$	$\mu_{hl} = \frac{b p^{h-2}}{a^2}$	$\mu_{hl} = \frac{p^{h-l+1}}{a^2}$

To show that (c) holds, let $D_h = \sum_{l=1}^{h} \mu_{hl} \mathbf{p}_{hl} \mathbf{p}_{hl}^T$ for $h = 2, \ldots, r$. Then

$$D_2 = \text{diag}\left(2(1-b), 0, \ldots, 0, 2(1-b)/a^2\right),$$

$$D_h = \text{diag}\left((1-b)^{h-2}, b(1-b)^{h-3}, b(1-b)^{h-4}, \ldots, b(1-b)^0, 0, \ldots, 1/a^2\right)$$

for $h = 3, \ldots, r-1$, and

$$D_r = \text{diag}\left((1-b)^{r-2}/b, (1-b)^{r-3}, (1-b)^{r-4}, \ldots, (1-b)^0, 1/a^2\right).$$

From this it is easy to deduce that $\sum_{h=2}^{r} D_h = I_r$. The straightforward computations are left to the reader. $\square$

Observe that the proof of Theorem 3.6 provides no information on the order of this rank r completely positive matrix whose cp-rank is $r(r+1)/2$.

By the approach used in this section, if A is a rank r completely positive matrix, $A = W^T W$ where W is an $r \times n$ matrix, then cp-rank A is the minimal number of vectors $\mathbf{q}_1, \ldots, \mathbf{q}_k$ in $\mathcal{C}_W^*$ which satisfy (3.11). The next proposition shows that in one special case it is enough to consider only vectors $\mathbf{q}_1, \ldots, \mathbf{q}_k$ which lie in a certain finite set of rays. The proof uses arguments similar to the ones used in Lemma 3.2, and is left as an exercise.

Proposition 3.4 *Let A be a rank r doubly nonnegative $n \times n$ matrix, $A = W^T W$ where W is $r \times n$ and every $0 \neq \mathbf{x} \in \mathcal{C}_W^*$ has a positive last entry. Denote $K = \mathcal{C}_W^* \cap \{\mathbf{x} \in \mathbf{R}^r \mid x_r = 1\}$, and suppose that $\|\mathbf{x}\| \leq r$ for every $\mathbf{x} \in K$. Then*

(a) *If A is completely positive, then $K \cap \{\mathbf{x} \mid \|\mathbf{x}\| = r\}$ is a nonempty finite set.*

(b) *If $K \cap \{\mathbf{x} \mid \|\mathbf{x}\| = r\} = \{\mathbf{p}_1, \ldots, \mathbf{p}_k\}$, then A is completely positive if and only if there exist nonnegative numbers λ_i such that*

$$\sum_{i=1}^{k} \lambda_i \mathbf{p}_i \mathbf{p}_i^T = I_r \tag{3.18}$$

and

$$\text{cp-rank } A = \min_{(\lambda_1, \ldots, \lambda_k) \in \Lambda} \{\text{supp} \, (\lambda_1, \ldots, \lambda_k)\},$$

where Λ is the set of all nonnegative solutions to (3.18).

Exercises

3.10　Use Theorem 3.5 to find an upper bound on the cp-ranks of all $n \times n$ completely positive matrices, and Theorem 3.4 to find an upper bound on the cp-ranks of all $n \times n$ completely positive matrices whose graph is triangle free.

3.11　Prove Proposition 3.4.

Notes.　Theorem 3.4 is due to [Hannah and Laffey (1983)] and the rest of this section is based on [Barioli and Berman (2003)]. In [Barioli (2001b)] it is shown that the cp-ranks of rank 3 matrices with at least one off-diagonal zero are actually bounded by 4 (one less than the general bound $3(3+1)/2 - 1$).

3.3　Completely positive matrices of a given order

We are looking for an upper bound on the cp-ranks of completely positive matrices of order n. By Theorem 3.5, if A is an $n \times n$ completely positive matrix, then

$$\text{cp-rank } A \leq \frac{n(n+1)}{2} - 1. \tag{3.19}$$

For $n = 3$ this bound is 5, and for $n = 4$ this bound is 9. In both cases we have a better bound in Theorem 3.3: cp-rank $A \leq n$. So is there a better upper bound then (3.19) on the cp-ranks of $n \times n$ matrices? So far, there are only partial results. Some of these results led Drew, Johnson and Loewy to conjecture the following:

Conjecture 3.1 [The DJL Conjecture] *If A is an $n \times n$ completely positive matrix, $n \geq 4$, then*

$$\text{cp-rank } A \leq \frac{n^2}{4}. \tag{3.20}$$

The DJL conjecture may be rephrased in terms of graphs as follows: For each graph G on $n \geq 4$ vertices,

$$\text{cp-rank } G \leq \frac{n^2}{4}. \tag{3.21}$$

Most of the results in this section treat matrices with a given graph. In particular, the results imply that (3.21) holds for certain classes of graphs. One result shows (3.20) to hold for any matrix which can be transformed into a diagonally dominant matrix by a positive diagonal congruence. However, at the time this book is being written, there may be a first example to disprove the DJL conjecture. Some details are provided in the notes at the end of the section.

We first consider CP realizations of trees.

Theorem 3.7 *If A is a completely positive matrix, and $G(A)$ is a tree, then* $\text{cp-rank } A = \text{rank } A$.

Proof. If $G(A)$ has 2 vertices, this is already known by Theorem 3.1. Suppose the theorem holds for every matrix whose graph is a tree on $n-1$ vertices. If $G(A)$ is a tree on n vertices, we may assume that 1 is a vertex of degree 1 in $G(A)$. Then

$$A = \begin{pmatrix} a_{11} & a_{12} & 0 & \cdots & 0 \\ a_{12} & a_{22} & \cdots & \cdots & a_{2n} \\ 0 & \vdots & & & \vdots \\ \vdots & \vdots & & & \vdots \\ 0 & a_{2n} & \cdots & \cdots & a_{nn} \end{pmatrix},$$

and we may write

$$A = \left(\begin{pmatrix} a_{11} & a_{12} \\ a_{12} & a_{12}^2/a_{11} \end{pmatrix} \oplus 0_{n-2} \right) + (0 \oplus \hat{A}),$$

where $\hat{A}$ is the Schur complement $A/[1]$. The matrix $\hat{A}$ is an $(n-1) \times (n-1)$ completely positive matrix whose graph is a tree, and rank $\hat{A} = \text{rank } A - 1$.

By the induction hypothesis cp-rank $\hat{A}$ = rank $\hat{A}$. Thus,

$$\text{rank } A \leq \text{cp-rank } A \leq 1 + \text{cp-rank } \hat{A} = 1 + \text{rank } \hat{A} = \text{rank } A,$$

so cp-rank A = rank A.   □

Since every graph has a nonsingular CP matrix realization, Theorem 3.7 implies

Corollary 3.2 *If G is a tree on n vertices, then* cp-rank $G = n$.

Trees are not the only graphs with the property that every CP matrix realization of the graph has cp-rank equal to the rank. The question which graphs have this property will be discussed in Section 3.4. For now, here is one such example.

Example 3.3   If

$$A = \begin{pmatrix} a_{11} & a_{12} & 0 & 0 \\ a_{12} & a_{22} & a_{23} & a_{24} \\ 0 & a_{23} & a_{33} & a_{34} \\ 0 & a_{24} & a_{34} & a_{44} \end{pmatrix}$$

is completely positive and rank $A = 3$, then cp-rank A = rank A = 3. This implies that the graph G which consists of two blocks, a triangle and an edge, has the property that every completely positive matrix A with $G(A) = G$ satisfies cp-rank A = rank A. We leave the details of the proof as an exercise.

Next we consider CP matrix realizations of triangle free graphs.

Theorem 3.8 *If A is an $n \times n$ completely positive matrix and $G(A)$ is triangle free, then*

$$\text{cp-rank } A \leq \max\{n, |E(G(A))|\}.$$

Proof. It suffices to prove the theorem for irreducible matrices.

Consider first the case that the irreducible A is singular. By Theorem 2.7, $M(A)$ is positive semidefinite and, as in the proof of Theorem 2.6, this implies that for some positive diagonal D the matrix DAD is diagonally dominant. Hence we may assume that A itself is diagonally dominant, and since it is singular, this means that

$$a_{ii} = \sum_{\substack{j=1 \\ j \neq i}}^{n} a_{ij} \quad \text{for} \quad i = 1, \ldots, n.$$

Let $B_{ij} = a_{ij}(\mathbf{e}_i + \mathbf{e}_j)(\mathbf{e}_i + \mathbf{e}_j)^T$, $1 \leq i < j \leq n$. Then $A = \sum_{1 \leq i < j \leq n} B_{ij}$ is a rank 1 representation of A with $|E(G(A))|$ nonzero summands, hence

$$\text{cp-rank}\, A \leq |E(G(A))| \leq \max\{n, |E(G(A))|\}.$$

Next we consider the case that A is nonsingular and the connected graph $G(A)$ is not a tree. By applying a positive diagonal congruence we may assume that A is strictly diagonally dominant (see Theorem 1.16 and Remark 1.4). Since $G(A)$ is connected and not a tree, there exists an edge which may be removed without disconnecting $G(A)$. Using a permutation similarity we may assume that the entry corresponding to this edge is a_{12}. For each $x \geq 1$ let $\mathbf{b}(x)^T = \sqrt{a_{12}}(x, 1/x, 0, \ldots, 0)$, and $B(x) = \mathbf{b}(x)\mathbf{b}(x)^T$. The matrix $M(A - B(1))$ is strictly diagonally dominant, and hence nonsingular. The matrix $M(A - B(\sqrt{a_{11}/a_{12}}))$ is not positive semidefinite. Let x_0 be the smallest $x \geq 1$ such that $M(A - B(x))$ is singular. The matrix $\hat{A} = A - B(x_0)$ is a singular completely positive matrix whose graph is triangle free and connected. By the first part of the proof, cp-rank $\hat{A} \leq |E(G(\hat{A}))| = |E(G(A))| - 1$. Since $A = \hat{A} + B(x_0)$, we have

$$\text{cp-rank}\, A \leq |E(G(A))|.$$

Note that in this case $|E(G(A))| = \max\{n, |E(G(A))|\}$.

Finally, consider the case that A is nonsingular and $G(A)$ is a tree. In that case $\max\{n, |E(G(A))|\} = n = |E(G(A))| + 1$, and by Theorem 3.7, cp-rank $A = \text{rank}\, A = n$. $\qquad \square$

Remark 3.3 The maximal cliques in a triangle free graph G are the edges. Hence, by Remark 3.1, cp-rank $A \geq |E(G)|$ for every CP matrix realization of G. This, combined with the proof of Theorem 3.8, shows that if G is a triangle free graph which is not a tree, and A is a completely positive matrix such that $G(A) = G$, then cp-rank $A = |E(G)|$.

Corollary 3.3 *If G is a connected triangle free graph which is not a tree, then* cp-rank $G = |E(G)|$.

In the coming results we need the following lemma.

Lemma 3.3 *If v is a vertex of degree 1 or 2 in a graph G, then*

$$\text{cp-rank}\, G \le d(v) + \text{cp-rank}\, (G - v).$$

Proof. Suppose that G is a graph on vertices $\{1, \ldots, n\}$ and $v = 1$. We consider first the case that $d(v) = 2$, and assume that 2 and 3 are the vertices adjacent to $v = 1$.

Let A be a CP matrix realization of G, and let $A = \sum_{i=1}^m \mathbf{b}_i \mathbf{b}_i^T$ be any rank 1 representation of A. Denote

$$\Omega_1 = \{1 \le i \le m \mid 1 \in \text{supp}\, \mathbf{b}_i\} \quad , \quad \Omega_2 = \{1, \ldots, m\} \setminus \Omega_1$$
$$B = \sum_{i \in \Omega_1} \mathbf{b}_i \mathbf{b}_i^T \quad , \quad C = \sum_{i \in \Omega_2} \mathbf{b}_i \mathbf{b}_i^T.$$

Then $A = B + C$, $B = B' \oplus 0_{n-3}$ and $C = 0_1 \oplus C'$ are both completely positive, and $G(C') \subseteq G - v$. We may assume that B' is singular. Otherwise, let δ_0 be the maximal δ for which $B - \delta E_{33}$ is positive semidefinite. The matrix $B_1 = B - \delta_0 E_{33}$ is a completely positive matrix, and it is zero except for its 3×3 leading principal submatrix, which is singular. The matrix $C_1 = C + \delta_0 E_{33}$ is obviously also completely positive, and has the same graph as C. Hence we may replace B and C by B_1 and C_1, respectively. Since the 3×3 matrix B' is singular,

$$\text{cp-rank}\, A \le \text{cp-rank}\, B' + \text{cp-rank}\, C' \le 2 + \text{cp-rank}\, C' \le 2 + \text{cp-rank}\, (G - v).$$

This completes the proof in the case that $d(v) = 2$.

Suppose then that $d(v) = 1$ and $v = 1$ is adjacent to 2. Let A be a CP realization of G. By the same argument as in the proof of Theorem 3.7,

$$A = \left(\begin{pmatrix} a_{11} & a_{12} \\ a_{12} & a_{12}^2/a_{11} \end{pmatrix} \oplus 0_{n-2} \right) + (0 \oplus \hat{A}),$$

where $\hat{A}$ a completely positive matrix whose graph is $G - v$. The desired inequality easily follows. □

Theorem 3.9 *For every $n \ge 4$, cp-rank $T_n = 2n - 4$, and there exists a CP matrix realization A of T_n such that cp-rank $A = k$ if and only if $n - 2 \le k \le 2n - 4$.*

Proof. By Example 3.1 and Theorem 3.3, cp-rank $T_4 = 4$. The graph T_n has $n - 2$ vertices of degree 2. Using Lemma 3.3 successively $n - 4$ times, we find that for $n \geq 4$

$$\text{cp-rank}\, T_n \leq \text{cp-rank}\, T_4 + 2(n - 4) = 4 + 2(n - 4) = 2n - 4. \qquad (3.22)$$

This implies that cp-rank $A \leq 2n - 4$ for every CP matrix realization A of T_n. The inequality $n - 2 \leq$ cp-rank A holds since $c(T_n) = n - 2$.

We need to show that for every $n - 2 \leq k \leq 2n - 4$ there exists a matrix A such that $G(A) = T_n$ and cp-rank $A = k$. This will also imply that equality holds in (3.22). So let

$$A_{n,0} = \begin{pmatrix} (n-2)J_2 & J_{2 \times (n-2)} \\ J_{2 \times (n-2)}^T & I_{n-2} \end{pmatrix} = \begin{pmatrix} n-2 & n-2 & 1 & 1 & \cdots & 1 \\ n-2 & n-2 & 1 & 1 & \cdots & 1 \\ 1 & 1 & 1 & 0 & \cdots & 0 \\ 1 & 1 & 0 & 1 & \ddots & \vdots \\ \vdots & \vdots & \vdots & \ddots & \ddots & 0 \\ 1 & 1 & 0 & \cdots & 0 & 1 \end{pmatrix}.$$

Then $A_{n,0} = BB^T$, where B is the $n \times (n - 2)$ matrix whose j-th column is zero, except for 1 in positions $1, 2, j + 2$:

$$B = \begin{pmatrix} 1 & 1 & 1 & \cdots & 1 \\ 1 & 1 & 1 & \cdots & 1 \\ 1 & 0 & 0 & \cdots & 0 \\ 0 & 1 & 0 & & \vdots \\ \vdots & 0 & 1 & \ddots & \\ \vdots & & \ddots & \ddots & 0 \\ 0 & \cdots & & 0 & 1 \end{pmatrix}.$$

Hence A is completely positive and cp-rank $A_{n,0} = n - 2$. By Lemma 2.1, for every $0 < \alpha < n - 2$ the matrix $A_{n,\alpha} = A_{n,0} + \left(\begin{pmatrix} \alpha & -\alpha \\ -\alpha & \alpha \end{pmatrix} \oplus 0_{n-2} \right)$ is also completely positive. We leave it as an exercise to show that for $r = 1, \ldots, n - 2$ and $r - 1 < \alpha < r$ the cp-rank of cp-rank $A_{n,\alpha}$ is $r + (n - 2)$ (Exercise 3.13). $\qquad \square$

The results so far show that the DJL conjecture is true for matrices whose graph is either triangle free or T_n:

Corollary 3.4   *If G is a graph on $n \geq 4$ vertices which is either triangle free or T_n, then*

$$\text{cp-rank}\, G \leq \frac{n^2}{4}.$$

Proof.   If G is triangle free, then $|E(G)| \leq n^2/4$ by Theorem 1.24, and since $n \leq n^2/4$ for $n \geq 4$, the inequality follows from Theorem 3.8.

Since $2n - 4 \leq n^2/4$ for every n, we also have cp-rank $T_n \leq n^2/4$.   □

To prove the DJL conjecture for general completely positive graphs, we need the following lemma.

Lemma 3.4   *Let $G = G_1 \cup G_2$, where G_1 is a completely positive graph on vertices $\{1, \ldots, k\}$, and G_2 is a graph on vertices $\{k, \ldots, n\}$. If A is a completely positive matrix with $G(A) = G$, then*

$$A = (A_1 \oplus 0_{n-k}) + (0_{k-1} \oplus A_2),$$

where A_1 and A_2 are completely positive, $G(A_1) = G_1$, $G(A_2) = G_2$, A_1 is singular, and rank $A = $ rank $A_1 + $ rank A_2.

Proof.   Let $A = \sum_{i=1}^{m} \mathbf{b}_i \mathbf{b}_i^T$ be any rank 1 representation of A, and denote

$$\Omega_1 = \{1 \leq i \leq m \,|\, \text{supp}\, \mathbf{b}_i \subseteq \{1, \ldots, k\}\} \quad , \quad \Omega_2 = \{1, \ldots, m\} \setminus \Omega_1$$
$$B = \sum_{i \in \Omega_1} \mathbf{b}_i \mathbf{b}_i^T \quad , \quad C = \sum_{i \in \Omega_2} \mathbf{b}_i \mathbf{b}_i^T.$$

Then $B = B' \oplus 0_{n-k}$, $C = 0_{k-1} \oplus C'$, B' and C' are completely positive, $G(B') = G_1$ and $G(C') = G_2$. If $\mathbf{e}_k \notin \text{cs}\,(B)$, let $A_1 = B'$ and $A_2 = C'$. If $\mathbf{e}_k \in \text{cs}\,(B)$, let δ_0 be the maximal $\delta > 0$ for which $B - \delta \mathbf{e}_k \mathbf{e}_k^T$ is positive semidefinite. Then $\mathbf{e}_k \notin \text{cs}\,(B - \delta_0 \mathbf{e}_k \mathbf{e}_k^T)$ (see Exercise 1.18). $B - \delta_0 \mathbf{e}_k \mathbf{e}_k^T = A_1 \oplus 0_{n-k}$ is doubly nonnegative, and since $G(A_1) = G_1$ is completely positive graph, A_1 is completely positive. Clearly the matrix $C + \delta_0 \mathbf{e}_k \mathbf{e}_k^T = 0_{k-1} \oplus A_2$ is also completely positive, and

$$A = (A_1 \oplus 0_{n-k}) + (0_{k-1} \oplus A_2).$$

Since $\mathbf{e}_k \notin \text{cs}\,(A_1 \oplus 0_{n-k}) \cap \text{cs}\,(0_{k-1} \oplus A_2)$, rank $A = $ rank $A_1 + $ rank A_2.□

We can now prove

Theorem 3.10   *If G is a completely positive graph on n vertices, then for every CP matrix realization A of G,*

$$\text{cp-rank } A \leq \max\{|E(G)|, \text{rank } A\},$$

If $n \geq 4$, then also

$$\text{cp-rank } G \leq \frac{n^2}{4}.$$

Proof.   If $n \leq 3$, the first claim holds by Theorem 3.2. If G is tree on $n \geq 4$ vertices and A is a completely positive matrix with $G(A) = G$, then by Theorem 3.7

$$\text{cp-rank } A = \text{rank } A \leq \max\{|E(G)|, \text{rank } A\} \leq n \leq \frac{n^2}{4}.$$

Hence both claims are true in this case.

So we need to prove the two claims for any completely positive graph G on $n \geq 4$ vertices which is not a tree. In this case $\max\{|E(G)|, \text{rank } A\} = |E(G)|$. We therefore need to prove that if G is such a graph, then

$$\text{cp-rank } G \leq |E(G)| \quad \text{and} \quad \text{cp-rank } G \leq n^2/4.$$

It suffices to consider connected graphs (Exercise 3.15). We prove both claims by induction on the number of blocks.

If G has no cut vertex, then G is K_4 or a T_n or it is bipartite. The two desired inequalities follow from Theorems 3.3 and 3.9, Corollary 3.3 and Corollary 3.4.

Let G be a connected completely positive graph which is not a tree and has k blocks, and suppose both claims hold for any such graph with less than k blocks. Then $G = G_1 \cup G_2$, where G_1 and G_2 are connected completely positive graphs, and $G_1 \cap G_2$ consists of a single vertex. Since G is not a tree, at least one of the graphs G_1 and G_2 is not a tree. If both are not trees, then for every $i = 1, 2$, either G_i is a triangle, in which case $\text{cp-rank } G_i = 3 = |E(G_i)|$, or G_i has at least 4 vertices, in which case $\text{cp-rank } G_i \leq |E(G_i)|$ by the induction hypothesis. In any case,

$$\text{cp-rank } G \leq \text{cp-rank } G_1 + \text{cp-rank } G_2 \leq |E(G_1)| + |E(G_2)| = |E(G)|.$$

Now suppose one of these subgraphs, say G_1, is a tree. We may assume that the vertices of G_1 are $\{1, \ldots, k\}$. Let A be a completely positive matrix with $G(A) = G$. Then by Lemma 3.4 $A = (A_1 \oplus 0_{n-k}) + (0_{k-1} \oplus A_2)$, where

A_1 and A_2 are completely positive, $G(A_1) = G_1$ and $G(A_2) = G_2$, and A_1 is singular. Hence

$$\text{cp-rank } A \leq \text{cp-rank } A_1 + \text{cp-rank } A_2 = \text{rank } A_1 + \text{cp-rank } A_2.$$

Since A_1 is singular, $\text{rank } A_1 \leq |E(G_1)|$, and by the induction hypothesis (or the case $n = 3$), $\text{cp-rank } A_2 \leq |E(G_2)|$. Hence

$$\text{cp-rank } A \leq |E(G_1)| + |E(G_2)| = |E(G)|.$$

We now show that $\text{cp-rank } G \leq n^2/4$. Suppose $|V(G_i)| = n_i$, $i = 1, 2$. If both graphs have at least 4 vertices each, then

$$\text{cp-rank } G \leq \frac{n_1^2}{4} + \frac{n_2^2}{4} \leq \frac{(n_1 + n_2 - 1)^2}{4} = \frac{n^2}{4} \qquad (3.23)$$

by the induction hypothesis. If one of the graphs, say G_1, has less than 4 vertices, then by Lemma 3.4 and Theorem 3.2,

$$\text{cp-rank } G \leq (n_1 - 1) + \text{cp-rank } G_2.$$

Since $n_1 - 1 \leq n_1^2/4$, we get again (3.23) in the case that G_2 has 4 vertices or more. If G_2 also has less than 4 vertices, than $\text{cp-rank } G_2 \leq n_2$ and we get that $\text{cp-rank } G \leq (n_1 - 1) + n_2 = n \leq n^2/4$. □

Recall that for every graph G,

$$\min\{\text{cp-rank } A \mid A \text{ is a CP matrix realization of } G\} = c(G).$$

It is natural to ask whether every $c(G) \leq k \leq \text{cp-rank } G$ is the cp-rank of some CP matrix realization of G. In the case that G is completely positive the answer is yes. In other words, the set of cp-ranks of CP matrix realizations of G has no gaps.

Theorem 3.11 *If G is a completely positive graph and*

$$c(G) \leq k \leq \text{cp-rank } G,$$

then k is the cp-rank of some CP realization of G.

Observe that the theorem holds trivially for any graph G which is triangle free and not a tree, since $c(G) = \text{cp-rank } G = |E(G)|$ for such graphs. The proof for general completely positive graphs is outlined in Exercises 3.16–3.20. It is not known whether the theorem holds for other types of graphs.

Next we show that the DJL conjecture holds for every graph on 5 vertices which is not the complete graph. The following lemma is used repeatedly in the proof.

Lemma 3.5 *Let A be an $n \times n$ doubly nonnegative matrix. Suppose that for some $1 \le r \ne s \le n$ the support of row r is a subset of the support of row s. Let*

$$\alpha = \min \left\{ \frac{a_{sj}}{a_{rj}} \,\middle|\, a_{rj} > 0 \right\},$$

$S = I_n - \alpha E_{sr}$, *and* $\hat{A} = SAS^T$. *Then:*

(a) $\hat{A}$ *is doubly nonnegative and* rank $\hat{A}$ = rank A.
(b) *If $a_{ij} = 0$ for some $1 \le i, j \le n$, then $\hat{a}_{ij} = 0$*
(c) *There is at least one more zero off-diagonal entry in row s of $\hat{A}$, compared to row s of A.*

Proof. We may assume without loss of generality that $r = 1$ and $s = 2$, and

$$\alpha = \min \left\{ \frac{a_{2j}}{a_{1j}} \,\middle|\, a_{1j} > 0 \right\}.$$

Then $\alpha > 0$ and

$$\hat{A} = SAS^T = \begin{pmatrix} a_{11} & a_{12} - \alpha a_{11} & a_{13} & \cdots & a_{1n} \\ a_{12} - \alpha a_{11} & \beta & a_{23} - \alpha a_{13} & \cdots & a_{2n} - \alpha a_{1n} \\ a_{13} & a_{23} - \alpha a_{13} & a_{33} & \cdots & a_{3n} \\ \vdots & \vdots & \vdots & & \vdots \\ a_{1n} & a_{2n} - \alpha a_{1n} & a_{3n} & \cdots & a_{nn} \end{pmatrix}$$

where $\beta = a_{22} - \alpha a_{12} - \alpha(a_{12} - \alpha a_{11})$. From this (and the assumption on the supports of rows 1 and 2) it is obvious that (b) holds. The matrix $\hat{A}$ is positive semidefinite, hence $\beta \ge 0$. By the choice of α, all other entries of $\hat{A}$ are also nonnegative, hence $\hat{A}$ is doubly nonnegative. Since A is positive semidefinite, $a_{22}/a_{12} \ge a_{12}/a_{11}$, hence the minimum α is attained at some $j \ne 2$. For this j we have $a_{2j} > 0$ and $\hat{a}_{2j} = 0$. $\square$

Theorem 3.12 *For every graph G on 5 vertices which is not the complete graph,* cp-rank $G \le 6$.

Proof. Suppose A is a 5×5 completely positive matrix with $a_{15} = 0$. We need to show that cp-rank $A \leq 6$. If there is an additional zero entry in the first row of A, then the degree of the vertex $v = 1$ in $G(A)$ is at most 2. In that case, by Lemma 3.3,

$$\text{cp-rank } A \leq 2 + \text{cp-rank } (G(A) - v) \leq 2 + 4 = 6.$$

Similarly, cp-rank $A \leq 6$ if there is an additional zero entry in the fifth row of A. We therefore assume that all entries in the first and fifth row are nonzero, except for a_{15} and a_{51}. Let $A = \sum_{i=1}^{m} \mathbf{b}_i \mathbf{b}_i^T$ be any rank 1 representation of A. Let

$$\Omega_1 = \{1 \leq i \leq m \,|\, 1 \in \text{supp } \mathbf{b}_i\} \quad, \quad \Omega_2 = \{1, \ldots, m\} \setminus \Omega_1$$
$$B = \sum_{i \in \Omega_1} \mathbf{b}_i \mathbf{b}_i^T \quad, \quad C = \sum_{i \in \Omega_2} \mathbf{b}_i \mathbf{b}_i^T.$$

Then $B = B' \oplus 0$, and $C = 0 \oplus C'$, where B' and C' are both 4×4 completely positive matrices. All the entries in the first row of B' and all the entries in the last row of C' are positive. So

$$B = \begin{pmatrix} + & + & + & + & 0 \\ + & + & * & * & 0 \\ + & * & + & * & 0 \\ + & * & * & + & 0 \\ 0 & 0 & 0 & 0 & 0 \end{pmatrix} \quad, \quad C = \begin{pmatrix} 0 & 0 & 0 & 0 & 0 \\ 0 & + & * & * & + \\ 0 & * & + & * & + \\ 0 & * & * & + & + \\ 0 & + & + & + & + \end{pmatrix},$$

where each $+$ denotes a positive entry and each $*$ a nonnegative one. We show that it is enough to consider the case that $B[2,3,4]$ and $C[2,3,4]$ each have at least two zero entries above the diagonal. If, say, $B[2,3,4]$ is positive or has just one zero entry above the diagonal, then by permuting rows (and columns) $2, 3, 4$ of A, we may assume that the second row of B' is positive. The support of row 1 in B is then contained in that of row 2. Let

$$\alpha = \min \left\{ \frac{b_{2j}}{b_{1j}} \,\middle|\, 1 \leq j \leq 4 \right\}$$

and $S = I_5 - \alpha E_{21}$. By Lemma 3.5, $\hat{B} = SBS^T$ is a doubly nonnegative matrix. Since it has at most 4 nonzero rows, it is completely positive. At least one of the entries $\hat{b}_{12}, \hat{b}_{23}, \hat{b}_{24}$ is zero. Since row 1 in C is zero,

$SCS^T = C$ and

$$\hat{A} = SAS^T = SBS^T + SCS^T = \hat{B} + C.$$

In particular, $\hat{A}$ is completely positive and cp-rank $A \leq$ cp-rank $\hat{A}$ (see Proposition 3.3). If $\hat{b}_{12} = 0$, then $\hat{a}_{12} = 0$, and by Lemma 3.3 we deduce that cp-rank $\hat{A} \leq 6$. So we need to consider the case that (at least) one of the entries $\hat{b}_{23}, \hat{b}_{24}$ is zero. If only one of them is zero, we may assume by permutation similarity that it is $\hat{b}_{23}$. But then the support of row 1 in $\hat{B}$ is contained in that of row 4, and we may repeat the above argument with

$$\beta = \min\left\{ \left. \frac{\hat{b}_{4j}}{\hat{b}_{1j}} \right| 1 \leq j \leq 4 \right\}$$

and $\hat{S} = I_5 - \beta E_{41}$. We get a matrix $\tilde{B} = \hat{S}\hat{B}\hat{S}^T$. Again we may restrict our attention to the case that the additional zero entry in the fourth row of $\tilde{B}$ is not in the first column. That is, $\tilde{B}[2,3,4]$ has at least two zero nondiagonal entries, and $\tilde{A} = \hat{S}\hat{A}\hat{S}^T = \tilde{B} + C$ is a completely positive matrix, cp-rank $A \leq$ cp-rank $\hat{A} \leq$ cp-rank $\tilde{A}$. A similar argument applies to C: we may use the fifth row of C to generate a matrix $\hat{C}$ which has two zeros in its $\{2,3,4\}$ principal submatrix, without changing $\tilde{B}$, except possibly permuting rows and columns $2,3,4$.

So assume $A = B + C$, $B = B' \oplus 0$, $C = 0 \oplus C'$, and each of the matrices B and C has at least two zero entries above the diagonal in its principal submatrix on rows $2,3,4$. If the two zeros of B are in the same positions as those of C, then A has two zeros in one of the rows 2, 3 or 4. In this case we deduce from Lemma 3.3 that cp-rank $A \leq 6$. Thus there is one case left to consider: the case that only one of the zeros of B and one of the zeros of C coincide. By permuting rows and columns $2,3,4$ we may assume that

$$B = \begin{pmatrix} + & + & + & + & 0 \\ + & + & 0 & 0 & 0 \\ + & 0 & + & + & 0 \\ + & 0 & + & + & 0 \\ 0 & 0 & 0 & 0 & 0 \end{pmatrix} \quad, \quad C = \begin{pmatrix} 0 & 0 & 0 & 0 & 0 \\ 0 & + & + & 0 & + \\ 0 & + & + & 0 & + \\ 0 & 0 & 0 & + & + \\ 0 & + & + & + & + \end{pmatrix},$$

where each $+$ denotes a positive entry. If both B' and C' are singular, then by Example 3.3, cp-rank $B' \leq 3$ and cp-rank $C' \leq 3$, and cp-rank $A \leq 6$. Hence we assume that at least one of the matrices B' and C' is nonsingular.

By symmetry we may assume B' is. By applying a positive diagonal congruence on $A = B + C$ we may assume that $b_{34} = b_{33}$. Then $b_{44} > b_{33}$, and for each $0 \le \alpha < b_{33}$ the matrix $\begin{pmatrix} b_{33} - \alpha & b_{33} - \alpha \\ b_{33} - \alpha & b_{44} - \alpha \end{pmatrix}$ is positive definite and nonnegative. For each such α and nonnegative β let

$$B(\alpha, \beta) = \begin{pmatrix} b_{11} & b_{12} & b_{13} & b_{14} \\ b_{12} & b_{22} + \beta & 0 & 0 \\ b_{13} & 0 & b_{33} - \alpha & b_{33} - \alpha \\ b_{14} & 0 & b_{33} - \alpha & b_{44} - \alpha \end{pmatrix}$$

$$C(\alpha, \beta) = \begin{pmatrix} c_{22} - \beta & c_{23} & 0 & c_{25} \\ c_{23} & c_{33} + \alpha & \alpha & c_{35} \\ 0 & \alpha & c_{44} + \alpha & c_{45} \\ c_{25} & c_{35} & c_{45} & c_{55} \end{pmatrix} .$$

Note that $B(0,0) = B'$ and $C(0,0) = C'$, and for every (α, β)

$$A = B(\alpha, \beta) + C(\alpha, \beta).$$

Define

$$T = \left\{ (\alpha, \beta) \,\middle|\, \begin{array}{c} \alpha > 0 \,,\ \beta \ge 0,\ B(\alpha, \beta) \text{ is singular, and} \\ B(\alpha, \beta) \text{ and } C(\alpha, \beta) \text{ are doubly nonnegative} \end{array} \right\}.$$

The set T is not empty: The matrix $B(0,0)$ is nonnegative and positive definite, while $B(b_{33}, 0)$ is nonnegative and not positive semidefinite. Hence there exists $0 < \alpha_0 < b_{33}$ such that $B(\alpha_0, 0)$ is a singular doubly nonnegative matrix. Since $C(\alpha_0, 0)$ is clearly doubly nonnegative, $(\alpha_0, 0) \in T$.

Let

$$T' = \{\alpha \,|\, \alpha > 0 \text{ and there exists } \beta \ge 0 \text{ such that } (\alpha, \beta) \in T\}.$$

T' is a bounded nonempty subset of $\mathbf{R}$. Let

$$\varphi = \sup_{\alpha \in T'} \alpha.$$

Then $\varphi \ge \alpha_0$. Let $\{\alpha_j\}$ be a nondecreasing sequence of elements of T' which converges to φ, and $\{\beta_j\}$ a sequence of nonnegative numbers such that $(\alpha_j, \beta_j) \in T$. Since $\{\beta_j\}$ is bounded (e.g., by c_{22}), it has a converging subsequence. We may assume that $\{\beta_j\}$ itself converges to a limit ψ. Then the sequences $\{B(\alpha_j, \beta_j)\}$ and $\{C(\alpha_j, \beta_j)\}$ both converge, to $B(\varphi, \psi)$ and

$C(\varphi, \psi)$ respectively. The matrices $B(\varphi, \psi)$ and $C(\varphi, \psi)$ are clearly doubly nonnegative, and $B(\varphi, \psi)$ is singular.

We will now show that $C(\varphi, \psi)$ is also singular. If this is not the case, $C(\varphi, \psi)$ is positive definite, and therefore there exists $\delta > 0$ such that $C(\varphi, \psi + \delta)$ is positive semidefinite. Consider the equality

$$
\begin{aligned}
0 &= \det B(\varphi, \psi) \\
&= -b_{12} \det \begin{pmatrix} b_{12} & b_{13} & b_{14} \\ 0 & b_{33} - \varphi & b_{33} - \varphi \\ 0 & b_{33} - \varphi & b_{44} - \varphi \end{pmatrix} + (b_{22} + \psi) \det(B(\varphi, \psi)[1, 3, 4]) \\
&= -b_{12}^2 (b_{33} - \varphi)(b_{44} - b_{33}) + (b_{22} + \psi) \det(B(\varphi, \psi)[1, 3, 4]).
\end{aligned}
$$

It implies that $B(\varphi, \psi)[1, 3, 4]$ has positive determinant, hence

$$
\det B(\varphi, \psi + \delta) = \det B(\varphi, \psi) + \delta \det(B(\varphi, \psi)[1, 3, 4]) > 0,
$$

and the doubly nonnegative matrix $B(\varphi, \psi + \delta)$ is positive definite. On the other hand, $B(b_{33}, \psi + \delta)$ is nonnegative but clearly not positive semidefinite. Thus there exists an $\varepsilon > 0$ such that $B(\varphi + \varepsilon, \psi + \delta)$ is a singular doubly nonnegative matrix. Since $C(\varphi, \psi + \delta)$ is doubly nonnegative, $C(\varphi + \varepsilon, \psi + \delta)$ is a doubly nonnegative matrix too. But this implies that $\varphi + \varepsilon \in T'$, which contradicts our choice of φ. Hence $C(\varphi, \psi)$ is a singular matrix. We may therefore assume that $A = B + C$, where $B = B' \oplus 0$ and $C = 0 \oplus C'$, B' and C' are singular completely positive matrices, and

$$
B = \begin{pmatrix} + & + & + & + & 0 \\ + & + & 0 & 0 & 0 \\ + & 0 & + & + & 0 \\ + & 0 & + & + & 0 \\ 0 & 0 & 0 & 0 & 0 \end{pmatrix}, \quad C = \begin{pmatrix} 0 & 0 & 0 & 0 & 0 \\ 0 & + & + & 0 & + \\ 0 & + & + & + & + \\ 0 & 0 & + & + & + \\ 0 & + & + & + & + \end{pmatrix}.
$$

We use once more Lemma 3.5, this time for

$$
\alpha = \min \left\{ \frac{c_{3j}}{c_{5j}} \; \middle| \; 2 \le j \le 5 \right\},
$$

$S = I_5 - \alpha E_{35}$. Then $\hat{A} = SAS^T = B + SCS^T$ is a completely positive matrix and it suffices to show that cp-rank $\hat{A} \le 6$. Let $\hat{C} = SCS^T$. If $\hat{c}_{23} = 0$ or $\hat{c}_{35} = 0$, then $G(\hat{A})$ has a vertex of degree at most 2, and we deduce by Lemma 3.3 that cp-rank $\hat{A} \le 6$. Otherwise, we have $\hat{A} = B + \hat{C}$, $B = B' \oplus 0$, $\hat{C} = 0 \oplus \hat{C}'$, B' and $\hat{C}'$ are both singular 4×4 completely positive matrices, and the graph of each of them has two blocks, a triangle and a single edge. By Example 3.3 we have cp-rank B = cp-rank B' = rank $B' \leq 3$ and cp-rank $\hat{C}$ = cp-rank $\hat{C}'$ = rank $\hat{C}' \leq 3$, which implies

$$\text{cp-rank } A \leq \text{cp-rank } \hat{A} \leq \text{cp-rank } B + \text{cp-rank } \hat{C} \leq 3 + 3 = 6.$$

$\square$

The last theorem in this section proves the DJL conjecture for completely positive matrices with a positive semidefinite comparison matrix and any graph. The proof relies on the following technical lemma.

Lemma 3.6   *Let A be an $n \times n$ diagonally dominant symmetric nonnegative matrix. Let r, s, t be distinct indices, $1 \leq r, s, t \leq n$. For $x > 0$ define $\mathbf{b}(x) \in \mathbf{R}^n$ by*

$$\mathbf{b}(x)_r = \sqrt{x}, \quad \mathbf{b}(x)_s = \frac{a_{rs}}{\sqrt{x}}, \quad \mathbf{b}(x)_t = \frac{a_{rt}}{\sqrt{x}}, \quad \mathbf{b}(x)_i = 0, \quad i \neq r, s, t.$$

If

$$a_{rs}, a_{rt} \leq x \leq a_{rr} \tag{3.24}$$

and

$$a_{st} x \geq a_{rs} a_{rt}, \tag{3.25}$$

then

(a) *$A(x) = A - \mathbf{b}(x)\mathbf{b}(x)^T$ is symmetric and nonnegative, and has diagonal dominance in all rows except possibly the r-th row.*
(b) *If in addition to (3.24) and (3.25),*

$$a_{rs} a_{rt} > 0, \tag{3.26}$$

then the diagonal dominance in rows s and t is strict.
(c) *If*

$$x = a_{rs} + a_{rt} \text{ and } (3.25) \text{ holds}, \tag{3.27}$$

then there is also a diagonal dominance in the r-th row, that is, A is a diagonally dominant symmetric nonnegative matrix.

Proof. We may assume that $r = 1, s = 2, t = 3$. Then the matrix $A(x)$ coincides with A in rows $4, \ldots, n$, and

$$A(x)[1,2,3] = \begin{pmatrix} a_{11} - x & 0 & 0 \\ 0 & (a_{22}x - a_{12}^2)/x & (a_{23}x - a_{12}a_{13})/x \\ 0 & (a_{23}x - a_{12}a_{13})/x & (a_{33}x - a_{13}^2)/x \end{pmatrix}$$

The nonnegativity of $A(x)$ in rows $4, \ldots, n$ and the diagonal dominance in these rows follow from the properties of A. The nonnegativity in rows $1, 2, 3$ is guaranteed by (3.24) and (3.25). If $i = 2$ or $i = 3$, diagonal dominance in row i is equivalent to

$$x \left(a_{ii} - \sum_{j \neq i} a_{ij} \right) + (x - a_{1i})a_{1i} + a_{12}a_{13} \geq 0,$$

and this inequality follows from (3.24) and the diagonal dominance of A. If (3.26) holds, the last inequality is actually a strict inequality and there is a strict diagonal dominance in rows $i = 2, 3$. Finally, if x is as in (3.27), then

$$(a_{11} - x) - \sum_{j=4}^{n} a_{1j} = a_{11} - \sum_{j=2}^{n} a_{1j} \geq 0,$$

and there is diagonal dominance in row 1 also. $\qquad\square$

Theorem 3.13 *If A is an $n \times n$ doubly nonnegative matrix, $n \geq 4$, with a positive semidefinite comparison matrix, then*

$$\text{cp-rank}\, A \leq \frac{n^2}{4}.$$

Moreover, A has a rank 1 representation of support 3 with at most $\lfloor n^2/4 \rfloor$ summands.

Proof. We we may assume that A is a diagonally dominant, symmetric nonnegative matrix (see the proof of Theorem 2.6). We prove the theorem by induction on n.

First, the case $n = 4$. Let A be a diagonally dominant completely positive matrix. By Theorem 3.3, cp-rank $A \leq 4$, so $A = \sum_{i=1}^{4} \mathbf{b}_i \mathbf{b}_i^T$, where $\mathbf{b}_i$ are nonnegative vectors. We want to show that there exists such a representation in which no vector is positive. If A has a zero entry this is clear. So we need only consider the case that A is positive. Without loss of generality we may assume that $a_{12} = \min a_{ij}$. Every $a_{12}, a_{13} \leq x \leq a_{11}$ satisfies (3.24), (3.25) and (3.26) (with $r = 1, s = 2, t = 3$). For such x, the matrix $A(x)$ is nonnegative, symmetric and has strict diagonal dominance in rows 2 and 3, and since $a_{14} > 0$, $A(x)[2, 3, 4]$ is strictly diagonally dominant. The matrix $A(a_{11})$ is not positive semidefinite, hence $\det(A(a_{11})) < 0$. On the other hand, by Lemma 3.6(c), $A(a_{12} + a_{13})$ is diagonally dominant. It is therefore positive semidefinite, and $\det(A(a_{12} + a_{13})) \geq 0$. Let x_0 be the smallest $x \geq a_{12} + a_{13}$ for which $\det(A(x)) = 0$. Then $A(x_0)$ is a rank 3 doubly nonnegative matrix, and $A(x_0)_{12} = A(x_0)_{13} = 0$. By Example 3.3, cp-rank $A(x_0) = \mathrm{rank}\, A(x_0) = 3$. Because of the zero entries of $A(x_0)$, if $A(x_0) = \sum_{i=1}^{3} \mathbf{b}_i \mathbf{b}_i^T$ is a rank 1 representation of $A(x_0)$, then $|\mathrm{supp}\,(\mathbf{b}_i)| \leq 3$ for each $1 \leq i \leq 3$. Hence $A = \sum_{i=1}^{3} \mathbf{b}_i \mathbf{b}_i^T + \mathbf{b}(x_0)\mathbf{b}(x_0)^T$ is a rank 1 representation of A in which each of the four vectors has support of size at most 3.

Now let $n \geq 5$, and assume that the theorem is true for $4 \leq m < n$. Let A be an $n \times n$ diagonally dominant completely positive matrix. By continuity we may assume that A is positive. We will show that such A can be represented as:

$$A = \mathbf{b}_1 \mathbf{b}_1^T + \ldots + \mathbf{b}_k \mathbf{b}_k^T + \tilde{A},$$

where $k = \lfloor n/2 \rfloor$, the vectors $\mathbf{b}_i$ are all nonnegative with $|\mathrm{supp}\,(\mathbf{b}_i)| \leq 3$, and the matrix $\tilde{A}$ is a diagonally dominant completely positive matrix with a zero row. By the induction hypothesis, $\tilde{A}$ has a rank 1 representation with $\lfloor (n-1)^2/4 \rfloor$ summands, each with support of size at most 3. This will imply that A has such a representation with $\lfloor (n-1)^2/4 \rfloor + \lfloor n/2 \rfloor = \lfloor n^2/4 \rfloor$ summands.

Consider two cases: n is even, or n is odd.

If $n = 2k$ is even, we may assume by permutation similarity that

$$\max_{1 \leq i \neq j \leq n} a_{ij} = a_{n-1\,n}$$

$$\max_{1 \leq i \neq j \leq n-2} a_{ij} = a_{n-3\,n-2}$$

$$\vdots$$

$$\max_{1 \leq i \neq j \leq 4} a_{ij} = a_{34}.$$

Apply Lemma 3.6 $k-1$ times, each time with $x = a_{rs} + a_{rt}$. The first time apply it to A, with $r = 1, s = n-1, t = n, x_1 = a_{1\,n-1} + a_{1n}$. Let $\mathbf{b}_1 = \mathbf{b}(x_1)$

and $A_1 = A - \mathbf{b}_1 \mathbf{b}_1^T$. By the lemma, A_1 is a diagonally dominant completely positive matrix. Now apply the lemma to A_1, with $r = 1$, $s = n - 3$, $t = n - 2$, $x_2 = a_{1\,n-3} + a_{1\,n-2}$. Let $\mathbf{b}_2 = \mathbf{b}(x_2)$, and $A_2 = A_1 - \mathbf{b}_2 \mathbf{b}_2^T$. Continue until the lemma is applied to A_{k-2}, with $r = 1$, $s = 3$, $t = 4$ and $x_{k-1} = a_{13} + a_{14}$. Denote $\mathbf{b}_{k-1} = \mathbf{b}(x_{k-1})$, and $A_{k-1} = A - \sum_{i=1}^{k-1} \mathbf{b}_i \mathbf{b}_i^T$. Finally, denote $a = (A_{k-1})_{11}$, and $\mathbf{b}_k^T = (\sqrt{a}, a_{12}/\sqrt{a}, 0, \ldots, 0)$. The matrix $\tilde{A} = A_{k-1} - \mathbf{b}_k \mathbf{b}_k^T$ is a diagonally dominant completely positive matrix with a zero first row, and $A = \mathbf{b}_1 \mathbf{b}_1^T + \ldots + \mathbf{b}_k \mathbf{b}_k^T + \tilde{A}$. We have $k = n/2$ and the vectors $\mathbf{b}_i$ are all nonnegative with $|\text{supp}\,(\mathbf{b}_i)| \leq 3$.

If $n = 2k + 1$ is odd, we may assume by permutation similarity that

$$\max_{1 \leq i \neq j \leq n} a_{ij} = a_{n-1\,n}$$
$$\max_{1 \leq i \neq j \leq n-2} a_{ij} = a_{n-3\,n-2}$$
$$\vdots$$
$$\max_{1 \leq i \neq j \leq 5} a_{ij} = a_{45}.$$

The matrix

$$\begin{pmatrix} a_{11} - \sum_{j=4}^{n} a_{1j} & a_{12} & a_{13} \\ a_{12} & a_{22} - \sum_{j=4}^{n} a_{2j} & a_{23} \\ a_{13} & a_{23} & a_{33} \end{pmatrix}$$

is completely positive, so by Lemma 2.7 either

$$\left(a_{11} - \sum_{j=4}^{n} a_{1j} \right) a_{23} \geq a_{12} a_{13}, \tag{3.28}$$

or

$$\left(a_{22} - \sum_{j=4}^{n} a_{2j} \right) a_{13} \geq a_{12} a_{23}. \tag{3.29}$$

Without loss of generality we may assume that (3.28) holds. As in the previous case, define $\mathbf{b}_1, \ldots, \mathbf{b}_{k-1}$, $A_1, \ldots, A_{k-1}$, using repeatedly Lemma 3.6. Then

$$A_{k-1}[1,2,3] = \begin{pmatrix} a_{11} - \sum_{j=4}^{n} a_{1j} & a_{12} & a_{13} \\ a_{12} & a_{22} & a_{23} \\ a_{13} & a_{23} & a_{33} \end{pmatrix}.$$

By (3.28) we may use Lemma 3.6 once more, with $r = 1$, $s = 2$, $t = 3$ and $x = a_{11} - \sum_{j=4}^{n} a_{1j}$. Denote $\mathbf{b}_k = \mathbf{b}(x)$, and $\tilde{A} = A - \mathbf{b}_k \mathbf{b}_k^T$. By the lemma, $\tilde{A}$ is a symmetric nonnegative matrix, with diagonal dominance in rows 2 to n. Since its first row is zero, it is a diagonally dominant completely positive matrix. We have $A = \mathbf{b}_1 \mathbf{b}_1^T + \ldots + \mathbf{b}_k \mathbf{b}_k^T + \tilde{A}$; $k = (n-1)/2$, and the vectors $\mathbf{b}_i$ are all nonnegative with $|\text{supp}(\mathbf{b}_i)| \leq 3$. $\qquad\square$

Observe that there exist $n \times n$ matrices with a rank 1 representation with at most $\lfloor n^2/4 \rfloor$ summands, but do not have a rank 1 representation of support 3 (see Exercise 3.22). Moreover,

Example 3.4 The matrix $A = J_5 + I_5$ has a rank 1 representation of support 3 and cp-rank $A = 5 < 5^2/4$. But A does not have a rank 1 representation with at most $\lfloor 5^2/4 \rfloor = 6$ summands which is at the same time also of support 3. We leave the first assertion as an exercise and prove the two others.

There are 10 different matrices obtained from $1/3 J_3 \oplus 0_2$ by permutation similarity. The matrix A is the sum of these 10 matrices.

Now suppose $A = \sum_{i=1}^{6} \mathbf{b}_i \mathbf{b}_i^T$ is a rank 1 representation of A of support 3. The matrix A has 20 nonzero off-diagonal elements, and each of the 6 matrices $\mathbf{b}_i \mathbf{b}_i^T$ has at most 6 nonzero off-diagonal entries. Therefore, there must exist $1 \leq r < s \leq 5$, and $1 \leq i \leq 6$ such that $\left(\mathbf{b}_i \mathbf{b}_i^T\right)_{rs} = a_{rs}$. We may assume that $r = 1$, $s = 2$, $i = 1$. Then $\mathbf{b}_1^T = (x, y, z, 0, 0)$ for some nonnegative x, y, z. The matrix

$$A' = A - \mathbf{b}_1 \mathbf{b}_1^T = \begin{pmatrix} 2 - x^2 & 1 - xy & 1 - xz & 1 & 1 \\ 1 - xy & 2 - y^2 & 1 - yz & 1 & 1 \\ 1 - xz & 1 - yz & 2 - z^2 & 1 & 1 \\ 1 & 1 & 1 & 2 & 1 \\ 1 & 1 & 1 & 1 & 2 \end{pmatrix}$$

is completely positive, and $xy = 1$. Then $\det(A'[1,2,4]) \leq 0$, with equality only if $x = y = 1$. But if $x = y = 1$, $\det(A'[1,2,4,5]) = -1$; a contradiction.

While all the evidence presented so far supports the DJL conjecture, it has recently been suggested that a counter example exists. See the notes at the end of the section.

Exercises

3.12   Prove the claims in Example 3.3.

3.13 Complete the proof of Theorem 3.9 by showing that for $r = 1, \ldots, n - 2$, if $r - 1 < \alpha < r$, then cp-rank $A_{n,\alpha} = r + n - 2$.
Hint: Show that if $A_{n,\alpha} = \sum \mathbf{b}_i \mathbf{b}_i^T$ is a rank 1 representation of $A_{n,\alpha}$, $r-1 < \alpha < r$, then at most $n-2-r$ of the vertices $3, \ldots, n$ may be in exactly one of the supports supp $\mathbf{b}_i$. Conclude that cp-rank $A_{n,\alpha} \geq 2r + (n-2-r)$. To prove the equality, consider first the case $r = n - 2$.

3.14 Show that the completely positive matrix $A_{n,\alpha}$ in the proof of Theorem 3.9 is critical for every $0 < \alpha < n - 2$.

3.15 Prove Theorem 3.10 for completely positive graphs which are not connected.

3.16 Say that a completely positive graph G is CP^1 if for every $c(G) \leq k \leq$ cp-rank G there exists a critical completely positive matrix A such that $G(A) = G$ and cp-rank $A = k$. By Exercise 3.6 K_4 is CP^1, and by Theorem 3.9 and Exercise 3.14 so is every T_n, $n \geq 4$. Show that every bipartite graph which is not a tree is also CP^1.

3.17 Say that a completely positive graph G is CP^2 if for every $c(G) \leq k \leq$ cp-rank $G - 1$ there exists a critical completely positive matrix A such that $G(A) = G$ and cp-rank $A = k$, and there exists no singular completely positive matrix such that $G(A) = G$ and cp-rank $A = $ cp-rank G. Show that K_3 is CP^2 and so is every tree.

3.18 Let $G = G_1 \cup G_2$, where G_1 and G_2 are connected subgraphs sharing exactly one vertex. Suppose $V(G_1) = \{1, \ldots, k\}$ and $V(G_2) = \{k, \ldots, n\}$. Show that if A_1 is a critical CP realization of G_1 and A_2 is a critical CP realization of G_2, then $A = (A_1 \oplus 0_{n-k}) + (0_{k-1} \oplus A_2)$ is a critical CP realization of G.

3.19 Suppose a connected completely positive graph G has a cut vertex, say $G = G_1 \cup G_2$, where G_1 and G_2 share exactly one vertex. Show that

(a) $c(G) = c(G_1) + c(G_2)$.
(b) If G_1 and G_2 are both CP^1, then

$$\text{cp-rank } G = \text{cp-rank } G_1 + \text{cp-rank } G_2$$

and G is CP^1.
(c) If G_1 is CP^2 and G_2 is CP^1, then

$$\text{cp-rank } G = \text{cp-rank } G_1 + \text{cp-rank } G_2 - 1$$

and G is CP^1.

(d) If G_1 is either K_2 or K_3 and G_2 is CP^2, then

$$\text{cp-rank}\, G = \text{cp-rank}\, G_1 + \text{cp-rank}\, G_2 - 1$$

and G is CP^2.

3.20   Prove Theorem 3.11. Show also that every completely positive graph is either CP^1 or CP^2.

3.21   Construct a completely positive matrix A which has a rank 1 representation of support 3 and at most $\lfloor n^2/4 \rfloor$ summands (n the order of A), such that $M(A)$ is not positive semidefinite.

3.22   Find an $n \times n$ completely positive matrix which has a rank 1 representation with no more than $\lfloor n^2/4 \rfloor$ summands, but does not have a rank 1 representation of support 3.

3.23   Complete Example 3.4 by showing that cp-rank $(J_5 + J_5) = 5$.

Notes.   Theorem 3.7 was proved in [Berman and Hershkowitz (1987)]. Theorem 3.8, and the conclusion that cp-rank $A \leq \lfloor n^2/4 \rfloor$ for every completely positive matrix of order $n \geq 4$ whose graph is triangle free, were proved in [Drew, Johnson and Loewy (1994)]. The fact that the bound was valid also in all other known cases led the authors to wonder whether this holds for every completely positive matrix of order $n \geq 4$ — hence the DJL conjecture, Conjecture 3.1. Theorem 3.8 improved known bounds: By [Kaykobad (1987)], cp-rank $A \leq |E(G(A))| + |V(G(A))|$ when A is diagonally dominant (see the proof of Theorem 2.5); In [Berman and Grone (1988)] it was shown that $|E(G(A))| \leq$ cp-rank $A \leq |E(G(A))| + 1$ when $G(A)$ is bipartite (see the algorithm in Section 2.5 for finding a rank 1 representation in this case).

It is shown in [Goldeberg (2003)] that if G is a completely positive graph on $n \geq 6$ vertices, then $|E(G)| \leq n^2/4$. Thus the second part of Theorem 3.10 follows from the first for $n \geq 6$.

Theorem 3.9 was proved in [Drew and Johnson (1996)]. The proof presented here is different. Prior to that paper, it was known that cp-rank $T_n \leq 2n-3$ by [Kogan and Berman (1993)]. The fact that every $n-2 \leq k \leq 2n-4$ is the cp-rank of some CP matrix realization of T_n is mentioned in [Drew and Johnson (1996)] and proved in [Zhang and Li (2002)], where it is part of the proof that there are no gaps in the set of cp-ranks of CP matrix realizations of a completely positive graph. The matrices $A_{n,\alpha}$ constructed here for the proof are different from the ones used in the original proof of [Zhang and Li (2002)].

Theorem 3.10 was proved in [Drew and Johnson (1996)]. Theorem 3.11 is taken from [Zhang and Li (2002)], and its proof as outlined in the exercises is based on the proof there.

Theorem 3.12 and its proof (including the preceding lemma) are taken from [Loewy and Tam (2003)], and Theorem 3.13 and Example 3.4 from [Berman and Shaked-Monderer (1998)].

F. Barioli has announced (in [Barioli (2002)]) an example of a 5×7 matrix W of rank 5 such that $A = W^T W$ is completely positive, $\|\mathbf{x}\| \leq 5$ for every $\mathbf{x} \in \mathcal{C}_W^*$ whose last entry is 1, and $\{\mathbf{x} \in \mathcal{C}_W^* \mid x_5 = 1$ and $\|\mathbf{x}\| = 5\} = \{\mathbf{p}_1, \ldots, \mathbf{p}_{14}\}$, where $\mathbf{p}_1 \mathbf{p}_1^T, \ldots, \mathbf{p}_{14} \mathbf{p}_{14}^T$ are linearly independent and $I_5 \in \text{cone}\,(\mathbf{p}_1 \mathbf{p}_1^T, \ldots, \mathbf{p}_{14} \mathbf{p}_{14}^T)$. By Proposition 3.4 this means that A is a 7×7 completely positive matrix with cp-rank $A = 14 > 12 = \lfloor 7^2/4 \rfloor$. Such A is a counter example to the DJL conjecture.

3.4   When is the cp-rank equal to the rank?

There is no complete answer to the question in the title of this section. The answer we present here is a qualitative one. We know that cp-rank $A =$ rank A if the completely positive matrix A is of rank 2 or of order 3 or less, and if $G(A)$ is a tree. Here is another example:

Example 3.5   By Remark 3.3, if A is a CP matrix realization of C_n, $n \geq 4$, then cp-rank $A = n$. If $n \geq 5$ is odd, then every CP realization of C_n is nonsingular by Remark 2.7. Hence cp-rank $A = $ rank A for every completely positive matrix A whose graph is an odd cycle of length 5 or more.

If $n \geq 4$ is even, there exist singular CP matrix realizations of C_n (see Exercise 2.29). If A is a such a matrix, then cp-rank $A = n > $ rank A.

Remark 3.4   A graph G on n vertices has the property that every non-singular matrix A with $G(A) = G$ satisfies cp-rank $A = $ rank A, if and only if cp-rank $G = n$. To see that, observe that if cp-rank $G = n$, then for every nonsingular matrix A with $G(A) = G$

$$n = \text{rank}\, A \leq \text{cp-rank}\, A \leq \text{cp-rank}\, G = n.$$

For the reverse implication, use the fact that every completely positive matrix A satisfying $G(A) = G$ is a limit of nonsingular completely positive matrices with the same graph: $A = \lim_{\varepsilon \to 0+} (A + \varepsilon I)$. If for each $\varepsilon > 0$ cp-rank $(A + \varepsilon I) = n$, then cp-rank $A \leq n$. Hence cp-rank $G = n$.

Of course, for every graph G on n vertices, cp-rank $G \geq n$, so the graphs which satisfy cp-rank $G = n$ attain the minimal possible graph cp-rank.

In this section we characterize all graphs G which satisfy

$$\text{For any nonsingular CP matrix realization } A \text{ of } G,$$
$$\text{cp-rank } A = \text{rank } A. \tag{3.30}$$

(By Remark 3.4 these are the graphs G for which cp-rank $G = |V(G)|$.) We will also characterize all graphs G which satisfy

$$\text{For any CP matrix realization } A \text{ of } G,$$
$$\text{cp-rank } A = \text{rank } A. \tag{3.31}$$

Trees, odd cycles and graphs on 3 vertices or less satisfy (3.31). Even cycles satisfy (3.30) but not (3.31). Every graph on 4 vertices satisfies (3.30). The graphs T_4 and K_4 do not satisfy (3.31) (see Example 3.1 and Exercise 3.6). Of course, every graph which satisfies (3.31), satisfies also (3.30). Some graphs satisfy neither:

Remark 3.5   A triangle free graph which has more edges than vertices does not satisfy (3.30). Every such graph G is not a tree, hence

$$\text{cp-rank } G = |E(G)| > |V(G)|.$$

A graph G satisfies (3.30) if and only if every component of G satisfies (3.30). The same holds for (3.31). We will therefore consider mostly connected graphs. In the next theorem we describe a class of connected graphs which satisfy (3.31).

Theorem 3.14   *If G is a block-clique graph, and each block of G is either an edge or a triangle, then G satisfies (3.31).*

Proof.   We use induction on the number of blocks in G. If G has no cut vertex, then G is either a K_2 or a K_3 and it satisfies (3.31). Suppose the claim holds for every block-clique graph which has less than m blocks, and let G be a graph with m blocks, $m \geq 2$, each of them a triangle or an edge. Then $G = G_1 \cup G_2$, where $G_1 \cap G_2$ is a single vertex, and G_1

and G_2 are connected. We may assume that $V(G_1) = \{1, \ldots, k\}$, and $V(G_2) = \{k, \ldots, n\}$. Let A be any CP matrix realization of G. Then, by Lemma 3.4 (since G_1 is a completely positive graph),

$$A = (A_1 \oplus 0_{n-k}) + (0_{k-1} \oplus A_2),$$

where A_1 and A_2 are completely positive, $G(A_1) = G_1$ and $G(A_2) = G_2$, A_1 is singular, and rank $A = $ rank $A_1 + $ rank A_2. By the induction hypothesis, cp-rank $A_1 = $ rank A_1 and cp-rank $A_2 = $ rank A_2. Therefore

$$
\begin{aligned}
\text{rank } A \;\; &\leq \;\; \text{cp-rank } A \\
&\leq \;\; \text{cp-rank } A_1 + \text{cp-rank } A_2 \\
&= \;\; \text{rank } A_1 + \text{rank } A_2 = \text{rank } A
\end{aligned}
$$

and this yields the desired equality. □

Theorem 3.14 implies the following generalization:

Corollary 3.5 *If H is a block of a connected graph G, and every other block of G is either an edge or a triangle, then G satisfies (3.30) ((3.31)) if and only if H satisfies (3.30) (respectively, (3.31)).*

Proof. Again we use induction on the number of blocks in G. If G has one block, than $G = H$ and the claim is obviously true. Suppose the claim holds for every connected graph which has less than m blocks, and let G be a graph with m blocks, $m \geq 2$, one of them H, and each of the others is either a triangle or an edge. Then $G = G_1 \cup G_2$, where $G_1 \cap G_2$ is a single vertex, G_1 and G_2 are connected. We may assume that $V(G_1) = \{1, \ldots, k\}$, and $V(G_2) = \{k, \ldots, n\}$. One of these subgraphs, say G_2, has H as one of its blocks. Then each of G_1's blocks is a K_2 or a K_3. In particular, G_1 is completely positive. Let A be a CP matrix realization of G. Then, by Lemma 3.4,

$$A = (A_1 \oplus 0_{n-k}) + (0_{k-1} \oplus A_2),$$

where A_1 and A_2 are completely positive, $G(A_1) = G_1$ and $G(A_2) = G_2$, A_1 is singular, and rank $A = $ rank $A_1 + $ rank A_2.

Now if H satisfies (3.31), then by the induction hypothesis G_2 also satisfies (3.31). Therefore

$$
\begin{aligned}
\text{rank } A \;\; &\leq \;\; \text{cp-rank } A \\
&\leq \;\; \text{cp-rank } A_1 + \text{cp-rank } A_2
\end{aligned}
$$

$$= \operatorname{rank} A_1 + \operatorname{rank} A_2 = \operatorname{rank} A.$$

That is, cp-rank $A = \operatorname{rank} A$.

If H satisfies (3.30), then by the induction hypothesis G_2 satisfies (3.30). If $\operatorname{rank} A = n$, the inequality $\operatorname{rank} A_1 \leq k - 1$, together with the equality $\operatorname{rank} A_1 + \operatorname{rank} A_2 = n$ implies that $\operatorname{rank} A_2 \geq n - k + 1$, and therefore $\operatorname{rank} A_2 = n - k + 1$. But $G(A_2) = G_2$, and G_2 satisfies (3.30), hence cp-rank $A_2 = \operatorname{rank} A_2$, and we deduce as above that cp-rank $A = \operatorname{rank} A = n$. $\qquad\square$

Theorem 3.15 *If G satisfies (3.30), then every subgraph of G also satisfies (3.30).*

Proof. First suppose that $V(H) = V(G) = \{1, \ldots, n\}$ and $E(H)$ is a proper subset of $E(G)$. Let A be a rank n CP matrix realization of H. For each edge $e \in E(G) \setminus E(H)$, denote by $\mathbf{1}_e$ the $(0, 1)$-vector of support e. For every $\varepsilon > 0$ let

$$A_\varepsilon = A + \sum_{e \in E(G) \setminus E(H)} \varepsilon \mathbf{1}_e \mathbf{1}_e^T.$$

Then A_ε is clearly completely positive, $G(A_\varepsilon) = G$ and $\operatorname{rank} A_\varepsilon = n$. Hence for each $\varepsilon > 0$ cp-rank $A_\varepsilon = n$. Since $A = \lim_{\varepsilon \to 0+} A_\varepsilon$, this implies that cp-rank $A \leq n$, and therefore cp-rank $A = n$.

Next suppose that $V(H)$ is a proper subset of $V(G)$. Without loss of generality assume that $V(H) = \{1, \ldots, k\}$ for some $k < n$. If A is a completely positive matrix such that $G(A) = H$ and $\operatorname{rank} A = k$, let

$$A_1 = A \oplus I_{n-k}.$$

Clearly A_1 is a completely positive matrix, $\operatorname{rank} A_1 = n$, $V(G(A_1)) = \{1, \ldots, n\}$, and $E(G(A_1))$ is a subset of $E(G)$. By the first part of the proof, cp-rank $A_1 = n$. It is easy to see that every minimal rank 1 representation of A_1 is of the form

$$\sum_{i=1}^{k} \mathbf{b}_i \mathbf{b}_i^T + \sum_{i=k+1}^{n} \mathbf{e}_i \mathbf{e}_i^T,$$

where $\sum_{i=1}^{k} \mathbf{b}_i \mathbf{b}_i^T$ is a rank 1 representation of A. Hence cp-rank $A = k$. $\square$

Theorem 3.15 together with Remark 3.5 imply:

Corollary 3.6 *A graph G which has a triangle free subgraph with more edges than vertices does not satisfy* (3.30).

We are going to show that the converse of Corollary 3.6 also holds. The proof requires a series of propositions. First we present some 2-connected graphs which satisfy (3.30), then we describe all the graphs which have no triangle free subgraphs with more edges than vertices. Using this description and Corollary 3.5 we will be able to show that all these graphs satisfy (3.30).

First — some additional 2-connected graphs which satisfy (3.30). For $n \geq 3$ denote by S_{2n} the graph obtained from C_{2n} by adding chords connecting each even vertex to the next even vertex (assuming the cycle vertices are numbered consecutively).

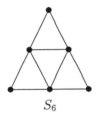
S_6

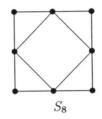
S_8

S_{16}

The chords generate an n-cycle (on the n even vertices). We will call this cycle the *inner cycle* of S_{2n}

We denote by S_5 the graph

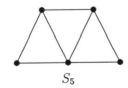
S_5

Proposition 3.5 *For every $n \geq 3$ the graph S_{2n} satisfies* (3.30).

Proof. We consider first the case $n \geq 4$. Renumber the vertices of S_{2n} so that 1 is a vertex of degree 1, adjacent to vertices 2 and 3. Let A be a completely positive matrix with $G(A) = S_{2n}$ and rank $A = 2n$. Take any rank 1 representation of A, $A = \sum_{i=1}^{m} \mathbf{b}_i \mathbf{b}_i^T$, and let

$$\Omega_1 = \{1 \le i \le m \,|\, \text{supp} \, \mathbf{b}_i \subseteq \{1,2,3\}\} \,, \quad \Omega_2 = \{1,2,\ldots,m\} \setminus \Omega_1,$$

$$B = \sum_{i \in \Omega_1} \mathbf{b}_i \mathbf{b}_i^T \,, \quad C = \sum_{i \in \Omega_2} \mathbf{b}_i \mathbf{b}_i^T.$$

Then B and C are completely positive, $B = B' \oplus 0_{2n-3}$, $C = 0_1 \oplus C'$. The graph $G(B')$ is a triangle, and $G(C')$ is a graph on $2n - 1$ vertices which is a "chain" of $n - 1$ triangles — every block of $G(C')$ is a K_3. Of course, $A = B + C$. Note that

$$B' = \begin{pmatrix} a_{11} & a_{12} & a_{13} \\ a_{12} & \alpha_0 & a_{23} \\ a_{13} & a_{23} & \beta_0 \end{pmatrix},$$

where $a_{12}^2/a_{11} \le \alpha_0 \le a_{22}$. If $\alpha_0 = a_{12}^2/a_{11}$, then $a_{23} = (a_{12}a_{13})/a_{11}$ and

$$B' = \begin{pmatrix} a_{11} & a_{12} & a_{13} \\ a_{12} & \alpha_0 & a_{23} \\ a_{13} & a_{23} & a_{13}^2/a_{11} \end{pmatrix} + \delta E_{33},$$

for some $\delta \ge 0$. In this case, denote $B'' = B' - \delta E_{33}$ and $C'' = C' + \delta E_{33}$. B'' is completely positive, and clearly so is C''; $A = (B'' \oplus 0_{2n-3}) + (0_1 \oplus C'')$ and

$$2n = \text{rank} \, A \le \text{cp-rank} \, A \le \text{cp-rank} \, B'' + \text{cp-rank} \, C'' \le 1 + (2n - 1) = 2n.$$

Now consider the case $a_{12}^2/a_{11} < \alpha_0 \le a_{22}$. For every $a_{12}^2/a_{11} < \alpha \le a_{22}$ denote

$$B'(\alpha) = \begin{pmatrix} a_{11} & a_{12} & a_{13} \\ a_{12} & \alpha & a_{23} \\ a_{13} & a_{23} & f(\alpha) \end{pmatrix},$$

where $f(\alpha)$ is the unique real number for which $B'(\alpha)$ is singular, i.e., $f(\alpha) = [a_{23}(a_{11}a_{23} - a_{12}a_{13}) - a_{13}(a_{12}a_{23} - \alpha a_{13})]/(a_{11}\alpha - a_{12}^2)$. Since $a_{11} > 0$ and $a_{11}\alpha - a_{12}^2 > 0$, $B'(\alpha)$ is a positive semidefinite matrix. In particular $f(\alpha) \ge 0$, so $B'(\alpha)$ is nonnegative, and therefore completely positive (Theorem 2.4). Let $B(\alpha) = B'(\alpha) \oplus 0_{2n-3}$, and $C(\alpha) = A - B(\alpha)$; $C(\alpha) = 0_1 \oplus C'(\alpha)$. Now if B', B, C' and C are as above, then $\beta_0 \ge f(\alpha_0)$, $B(\alpha_0) = B - (\beta_0 - f(\alpha_0))E_{33}$, and $C(\alpha_0) = C + (\beta_0 - f(\alpha_0))E_{33}$. Hence $C(\alpha_0)$ is completely positive. On the other hand, it is clear that $C(a_{22})$

is not positive semidefinite. By the continuity of the eigenvalues of $C(\alpha)$, it follows that there exists a $\alpha_0 < \alpha_1 < a_{22}$ such that $C(\alpha_1)$ is a singular positive semidefinite matrix. In particular, $C(\alpha_1)_{22} > 0$, $C(\alpha_1)_{33} > 0$, so $C(\alpha_1)$ is doubly nonnegative. The graph $G(C(\alpha_1))$ is completely positive, hence $C(\alpha_1)$ is a completely positive matrix. Thus we have

$$
\begin{aligned}
2n = \operatorname{rank} A \ &\leq\ \text{cp-rank } A \\
&\leq\ \text{cp-rank } B'(\alpha_1) + \text{cp-rank } C'(\alpha_1) \\
&\leq\ 2 + (2n - 2) = 2n.
\end{aligned}
$$

(The last inequality holds since both $B'(\alpha_1)$ and $C'(\alpha_1)$ are singular and their graphs satisfy (3.31) by Theorem 3.14.)

For the proof that S_6 also satisfies (3.30) we need to show first that S_5 satisfies (3.30). Number the vertices of S_5:

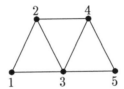

Let A be a completely positive matrix with $G(A) = S_5$ and rank $A = 5$. We show that cp-rank $A \leq 5$. Let $A = \sum_{i=1}^m \mathbf{b}_i \mathbf{b}_i^T$ be a rank 1 representation of A, and denote

$$
\Omega_1 = \{1 \leq i \leq m \mid 1 \in \operatorname{supp} \mathbf{b}_i\}\ ,\quad \Omega_2 = \{1, \ldots, m\} \setminus \Omega_1
$$
$$
A_1 = \sum_{i \in \Omega_1} \mathbf{b}_i \mathbf{b}_i^T\ ,\quad A_2 = \sum_{i \in \Omega_2} \mathbf{b}_i \mathbf{b}_i^T.
$$

Both matrices are completely positive and $A = A_1 + A_2$. Rows 4 and 5 of A_1 are zero, and $A_2 = 0_1 \oplus A_2'$. The support of row 5 in A_2 is contained in that of row 4. Let

$$
a = \min\left\{ \left. \frac{(A_2)_{4j}}{(A_2)_{5j}} \right| j = 3, 4, 5 \right\}
$$

and $S = I_5 - aE_{45}$. By Lemma 3.5, SA_2S^T is a doubly nonnegative matrix and at least one of its entries in positions 43 and 45 is zero. Since it has only four nonzero rows, it is completely positive. Row 4 of A_1 is zero, hence $SA_1S^T = A_1$ and $SAS^T = SA_1S^T + SA_2S^T = A_1 + SA_2S^T$ is completely positive. Also, rank (SAS^T) = rank A = 5, and cp-rank $A \leq$ cp-rank (SAS^T) (see Proposition 3.3). It therefore suffices to show that cp-rank (SAS^T) = 5. But the graph of SAS^T is contained in one of the following graphs:

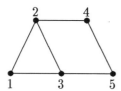   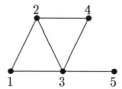

The graph on the left is a subgraph of S_8; the one on the right has two blocks, a K_2 and a T_4. By the beginning of this proof and Theorem 3.15, Theorem 3.3 and Corollary 3.5, both graphs satisfy (3.30), hence cp-rank $(SAS^T) \leq 5$.

We now show by similar arguments that S_6 also satisfies (3.30). Renumber the vertices:

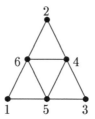

Let A be a rank 6 CP matrix realization of S_6. Let $A = \sum_{i=1}^m \mathbf{b}_i \mathbf{b}_i^T$ be a rank 1 representation of A and denote:

$$\Omega_1 = \{1 \leq i \leq m \,|\, 1 \in \text{supp}\,\mathbf{b}_i\}$$
$$\Omega_2 = \{1 \leq i \leq m \,|\, 2 \in \text{supp}\,\mathbf{b}_i\}$$
$$\Omega_3 = \{1, \ldots, m\} \setminus (\Omega_1 \cup \Omega_2).$$

(Note that $\Omega_1 \cap \Omega_2 = \emptyset$.) Let

$$A_1 = \sum_{i \in \Omega_1} \mathbf{b}_i \mathbf{b}_i^T \quad , \quad A_2 = \sum_{i \in \Omega_2} \mathbf{b}_i \mathbf{b}_i^T \quad , \quad A_3 = \sum_{i \in \Omega_3} \mathbf{b}_i \mathbf{b}_i^T.$$

These are three completely positive matrices, and $A = A_1 + A_2 + A_3$. Rows $2, 3, 4$ of A_1 are zero, rows $1, 3, 5$ of A_2 are zero, and rows 1 and 2 of A_3 are zero. In A_3, the support of row 3 is contained in that of row 4. Let

$$a = \min\left\{\frac{(A_3)_{4j}}{(A_3)_{3j}} \,\middle|\, j = 3, 4, 5\right\},$$

and $S = I_6 - aE_{43}$. Then SA_3S^T is a completely positive matrix, and at least one of its entries in positions 43 or 45 is zero;

$$SAS^T = SA_1S^T + SA_2S^T + SA_3S^T = A_1 + A_2 + SA_3S^T$$

is completely positive, and it suffices to prove that cp-rank $(SAS^T) \leq 6$. The graph of SAS^T is contained in one of the following graphs

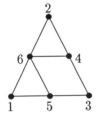
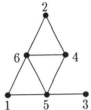

The graph on the left is a subgraph of S_8, and the graph on the right has two blocks: a K_2 and an S_5. By the beginning of this proof and Corollary 3.5 both graphs satisfy (3.30). Hence $G(SAS^T)$ satisfies (3.30) and cp-rank $(SAS^T) \leq 6$. $\qquad \square$

We turn to describe all blocks which contain no triangle free graph with more edges than vertices.

Proposition 3.6 *If G is a block which has no triangle free subgraph with more edges than vertices, then G is either a K_4 or a subgraph of S_{2n} for some $n \geq 3$.*

Proof. If G is a block on 2 vertices, then it is a K_2, if G is a block on 3 vertices, then it is a K_3. In both cases, G is a subgraph of S_6. We need to show that if G is a block on $n \geq 4$ vertices, which has no triangle free subgraph with more edges than vertices, then G is either a K_4 or a subgraph of S_{2n} for some $n \geq 3$. The only blocks on 4 vertices are C_4, T_4 and K_4. All three fit the description in the statement of the proposition.

Hence we consider the case $n \geq 5$. Let C be a cycle subgraph of G of maximal length. We first show that C is a spanning cycle of G. Suppose C is not a spanning cycle of G. Then C is a cycle on k vertices, $k < n$, and there exists a vertex v of G which is not a vertex of C. Since G is 2-connected, v lies on a path P connecting two vertices of C, v_1 and v_2, such that all the internal vertices of P are not vertices of C (see Exercise 1.30). Denote the length of P by s, then $s \geq 2$.

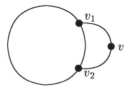

If v_1 and v_2 are adjacent in C, then the graph consisting of all edges of C other than $\{v_1, v_2\}$ and the edges of P is a cycle on $k - 1 + s \geq k + 1$ vertices, in contradiction to the choice of C. If v_1 and v_2 are not adjacent in C, then $k \geq 3$ and $C \cup P$ is a triangle free graph on $k + s - 1$ vertices, which has $k + s$ edges, and this contradicts the initial assumption regarding G. Hence there is no such vertex v, and the maximal cycle C is a spanning cycle of G.

We now consider which chords of C may be edges of G. There is no edge in G which is a chord between two vertices u, v such that the $d_C(u, v) \geq 3$, since the graph consisting of C and such a chord would be a triangle free subgraph of G which has n vertices and $n+1$ edges. Hence, if there exists an edge of G which is a chord of C, it would be some $\{u, v\}$ where $d_C(u, v) = 2$. If w is the vertex such that $d_C(u, w) = 1$ and $d_C(v, w) = 1$, then there is no edge $\{w, x\}$ of G which is a chord of C. Otherwise, $d_C(w, x) = 2$, so either u or v is halfway between w and x. Assume without loss of generality v is. Then the graph consisting of the two chords, and all of C's edges except for $\{w, v\}$, is a triangle free graph on n vertices with $2 + (n - 1) = n + 1$ edges. (The fact that $n \geq 5$ guarantees an additional vertex on the cycle C, between u and x.)

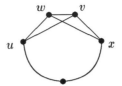

This completes the proof that G is a subgraph of some S_{2m}. □

Theorem 3.16 *If a connected graph G has no triangle free subgraph with more edges than vertices, then each block of G is a K_4 or a subgraph of S_{2n} for some $n \geq 3$, and at most one of the blocks of G has more than 3 vertices.*

Proof. If G has no triangle free subgraph with more edges than vertices, then no block of G has such subgraph. By Proposition 3.6 each block of G is either a K_4 or a subgraph of S_{2n} for some $n \geq 3$. Suppose two of these blocks have 4 vertices or more. By the proof of Proposition 3.6, each of these large blocks has a spanning cycle. Let C_1 and C_2 be spanning cycles of these two blocks. The cycles may share a vertex, or, if they don't, there is a path in G from a vertex of C_1 to a vertex of C_2 (whose internal vertices are not vertices of $C_1 \cup C_2$). A graph consisting of two cycles on 4 or more vertices sharing a vertex, or two such cycles and a path connecting them, is a triangle free graph with more edges than vertices. But this contradicts the assumption on G, hence there cannot be two blocks on 4 or more vertices in G. □

With Theorem 3.16, the characterization of all graphs which satisfy (3.30) is complete. The results are summarized in Theorem 3.17.

Theorem 3.17 *Let G be a connected graph, then the following are equivalent:*

(a) *For every nonsingular completely positive matrix realization A of G, cp-rank $A = \text{rank } A$.*
(b) *cp-rank $G = |V(G)|$.*
(c) *G contains no triangle free graph with more edges than vertices.*
(d) *Each block of G is either a K_4 or a subgraph of S_{2m} for some $m \geq 3$, and at most one of the blocks of G has more than 3 vertices.*

We now turn to characterize the graphs which satisfy (3.31). Since every such graph satisfies also (3.30), it suffices to check which of the graphs described in Theorem 3.17 satisfy (3.31).

Consider first 2-connected graphs which are subgraphs of S_{2m}, for some $m \geq 4$. Any such graph is a subgraph of some S_{2k}, $k \geq 4$, which contains the inner cycle C of S_{2k}. If it is not C itself, then it contains also $1 \leq r \leq k$ triangles, each consisting of an edge e of C, and a vertex not in C, which is joined by edges to e's ends. We start with one such example.

Example 3.6 Consider a 2-connected graph G which is a subgraph of $S_{2(n-1)}$, $n > 4$ even, containing the inner cycle and exactly one triangle. Let $V(G) = \{1, \ldots, n\}$ and suppose the edges are $\{i, i+1\}$, $i = 1, \ldots, n-1$, $\{1, n-1\}$ and $\{1, n\}$.

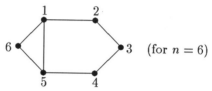

(for $n = 6$)

We show that G does not satisfy (3.31). Specifically, we construct a completely positive matrix A such that $G(A) = G$, rank $A = n - 1$ and cp-rank $A = n$. Let R be the following $n \times n$ matrix:

$$
R = \begin{pmatrix}
1 & 0 & 0 & \cdots & \cdots & 1 & 2 \\
1 & 1 & 0 & \cdots & \cdots & 0 & 0 \\
0 & 1 & 1 & & & \vdots & \vdots \\
0 & 0 & 1 & \ddots & & \vdots & \vdots \\
& & & \ddots & \ddots & & \\
0 & 0 & 0 & & 1 & 1 & 0 \\
0 & 0 & 0 & \cdots & 0 & 1 & 1
\end{pmatrix},
$$

and let $A = RR^T$. Then

$$
A = \begin{pmatrix}
6 & 1 & 0 & \cdots & \cdots & 1 & 3 \\
1 & 2 & 1 & & \cdots & 0 & 0 \\
0 & 1 & 2 & 1 & & \vdots & \vdots \\
\vdots & & \ddots & \ddots & \ddots & & \vdots \\
& & & \ddots & 2 & 1 & 0 \\
1 & 0 & \cdots & & 1 & 2 & 1 \\
3 & 0 & \cdots & \cdots & 0 & 1 & 2
\end{pmatrix}.
$$

Clearly, A is completely positive, $G(A) = G$, and cp-rank $A \leq n$. It is also easy to see that rank $A = n - 1$: Denote the i-th column of R by R_i, then the first $n - 1$ columns of R are linearly independent and $R_n = \sum_{i=1}^{n-1}(-1)^{i+1}R_i$. We show that cp-rank $A \geq n$. Let

$$A = \sum_{i=1}^{m} \mathbf{b}_i \mathbf{b}_i^T$$

be a minimal rank 1 representation of A. Let

$$\Omega_1 = \{1 \leq i \leq m \mid n \in \operatorname{supp} \mathbf{b}_i\}, \quad \Omega_2 = \{1, \ldots, m\} \setminus \Omega_1$$
$$B = \sum_{i \in \Omega_1} \mathbf{b}_i \mathbf{b}_i^T, \quad C = \sum_{i \in \Omega_2} \mathbf{b}_i \mathbf{b}_i^T.$$

Then B and C are completely positive. In B only rows $1, n - 1, n$ are nonzero, and the matrix C has the form $0_1 \oplus C'$. By the minimality of the representation, cp-rank $B = |\Omega_1|$, cp-rank $C = $ cp-rank $C' = |\Omega_2|$, and cp-rank $A = $ cp-rank $B + $ cp-rank C. If $|\Omega_1| = 1$, then

$$B[1, n - 1, n] = \begin{pmatrix} \frac{9}{2} & \frac{3}{2} & 3 \\ \frac{3}{2} & \frac{1}{2} & 1 \\ 3 & 1 & 2 \end{pmatrix},$$

but this would imply that the $1\,(n-1)$ element of $C = A - B$ is $-1/2$, which is impossible since C is completely positive. Hence $|\Omega_1| \geq 2$, which implies that cp-rank $B \geq 2$. Since $G(C')$ is either the cycle on $n - 1$ vertices or the path from vertex 1 to vertex $n - 1$, $c(G') \geq n - 2$. Thus cp-rank $C' \geq n - 2$, and we get

$$\text{cp-rank } A = \text{cp-rank } B + \text{cp-rank } C \geq 2 + (n - 2) = n.$$

We use Example 3.6 to prove that any block which is a subgraph of S_{2k}, $k \geq 4$, and is not an edge or a triangle, does not satisfy (3.31).

Proposition 3.7 *If $G \subseteq S_{2k}$, $k \geq 4$, is 2-connected which is not an odd cycle, then G does not satisfy* (3.31).

Proof. We may assume that G consists of the inner cycle of S_{2k} and r triangles, $r \geq 0$; G has $n = k + r$ vertices. We prove by induction on r that there exists a completely positive matrix A such that $G(A) = G$, rank $A = n - 1$, and cp-rank $A = n$. By Example 3.5, if k is even and $r = 0$, there exists a completely positive matrix A such that $G(A) = G$, rank $A = n-1$, cp-rank $A = n$. By Example 3.6, this is true also in the case that k is odd and $r = 1$. For the induction step we assume that the claim holds for any 2-connected subgraphs of S_{2k} which contains the inner cycle and $r - 1$ triangles. Let $G \subseteq S_{2k}$ be a 2-connected graph with r triangles. Then G has $n = k + r$ vertices, and we may assume that the vertex 1 is not a vertex of the inner cycle and that it is adjacent to the vertices 2 and 3 which are adjacent vertices of the inner cycle. Denote by G' the graph obtained from G by deleting vertex 1 and the two edges incident with it. By the induction hypothesis there exists a completely positive matrix A' such that $G(A') = G'$, rank $A' = n - 2$ and cp-rank $A' = n - 1$. Let

$$A = (J_3 \oplus 0_{n-2}) + (0_1 \oplus A').$$

The matrix A is clearly completely positive, $G(A) = G$, and rank $A =$ rank $A' + 1 = n - 1$. We show that cp-rank $A = $ cp-rank $A' + 1 = n$. Let

$$A = \sum_{i=1}^{m} \mathbf{b}_i \mathbf{b}_i^T$$

be a minimal rank 1 representation of A. Let

$$\Omega_1 = \{1 \le i \le m \,|\, \operatorname{supp} \mathbf{b}_i \subseteq \{1, 2, 3\}\} \quad , \quad \Omega_2 = \{1, \ldots, m\} \setminus \Omega_1,$$
$$B = \sum_{i \in \Omega_1} \mathbf{b}_i \mathbf{b}_i^T \quad , \quad C = \sum_{i \in \Omega_2} \mathbf{b}_i \mathbf{b}_i^T.$$

Then $B = B' \oplus 0_{n-3}$, $C = 0_1 \oplus C'$, $A = B + C$, and by the minimality of the representation cp-rank $A = $ cp-rank $B + $ cp-rank C. Note that

$$B' = \begin{pmatrix} 1 & 1 & 1 \\ 1 & + & \alpha + 1 \\ 1 & \alpha + 1 & + \end{pmatrix}, \tag{3.32}$$

where $+$ denotes a positive entry and $\alpha > 0$ is the 23 entry of $0_1 \oplus A'$. We may write $B' = J_3 + (0_1 \oplus B'')$, where B'' is the Schur complement $B'/[1]$. Then rank $B' = $ rank $B'' + 1$, and by (3.32) B'' is nonnegative. Since both B' and B'' are of order ≤ 3, the rank equality implies also that cp-rank $B' = $ cp-rank $B'' + 1$. The equality

$$A = (J_3 \oplus 0_{n-3}) + [C + (0_1 \oplus B'' \oplus 0_{n-3})].$$

implies that $C + (0_1 \oplus B'' \oplus 0_{n-3}) = 0_1 \oplus A'$, and therefore

$$\text{cp-rank } A' \le \text{cp-rank } C + \text{cp-rank } B''.$$

But then

$$\begin{aligned}
\text{cp-rank } C + \text{cp-rank } B'' + 1 &= \text{cp-rank } A \\
&\leq 1 + \text{cp-rank } A' \\
&\leq 1 + \text{cp-rank } C + \text{cp-rank } B''.
\end{aligned}$$

This implies that cp-rank A = cp-rank $A' + 1 = n$. $\qquad\square$

Blocks which are subgraphs of S_6, and are neither an edge nor a triangle, also do not satisfy (3.31).

Proposition 3.8 *If $G \subseteq S_6$ is 2-connected, and is not an odd cycle, then G does not satisfy (3.31).*

Proof. Each such proper subgraph of S_6 is either a T_4, or a C_4, or a subgraph of S_8, or S_5. By Examples 3.1 and 3.5, and Proposition 3.7, we only have to show that S_5 and S_6 do not satisfy (3.31).

Recall that by Example 3.1,

$$A_1 = \begin{pmatrix} 6 & 3 & 3 & 0 \\ 3 & 5 & 1 & 3 \\ 3 & 1 & 5 & 3 \\ 0 & 3 & 3 & 6 \end{pmatrix}$$

is a CP matrix realization of T_4, rank $A_1 = 3$, and cp-rank $A_1 = 4$. Let $A_2 = (J_3 \oplus 0_2) + (0 \oplus A_1)$. The matrix A_2 is a CP matrix realization of S_5 and rank $A_2 = 4$. Let $A_2 = \sum_{i=1}^m \mathbf{b}_i \mathbf{b}_i^T$ be a minimal rank 1 representation of A_2. If $1 \in \operatorname{supp} \mathbf{b}_i$ for exactly one i, say $i = 1$, then necessarily $\mathbf{b}_1 \mathbf{b}_1^T = J_3 \oplus 0_2$, and cp-rank A_2 = cp-rank $\mathbf{b}_1 \mathbf{b}_1^T$ + cp-rank $A_1 = 1 + 4 = 5$. Suppose 1 belongs to two of the supports, say $\operatorname{supp} \mathbf{b}_1$ and $\operatorname{supp} \mathbf{b}_2$. We argue as in Example 3.1 that 5 also belongs to two supports, say $\operatorname{supp} \mathbf{b}_3, \operatorname{supp} \mathbf{b}_4$. But if $1 \in \operatorname{supp} \mathbf{b}_i$ or $5 \in \operatorname{supp} \mathbf{b}_i$, then $\{2,4\} \not\subseteq \operatorname{supp} \mathbf{b}_i$. Thus there is a fifth vector $\mathbf{b}_5$ such that $\{2,4\} \subseteq \operatorname{supp} \mathbf{b}_5$. Hence cp-rank $A_2 = m \geq 5$ and S_5 does not satisfy (3.31) . By a similar argument,

$$A_3 = (A_1 \oplus 0_1) + \begin{pmatrix} 1 & 0 & 1 & 0 & 1 \\ 0 & 0 & 0 & 0 & 0 \\ 1 & 0 & 1 & 0 & 1 \\ 0 & 0 & 0 & 0 & 0 \\ 1 & 0 & 1 & 0 & 1 \end{pmatrix} = \begin{pmatrix} 7 & 3 & 4 & 0 & 1 \\ 3 & 5 & 1 & 3 & 0 \\ 4 & 1 & 6 & 3 & 1 \\ 0 & 3 & 3 & 6 & 0 \\ 1 & 0 & 1 & 0 & 1 \end{pmatrix}$$

is also completely positive, $G(A_3) = S_5$, rank $A_3 = 4$, and cp-rank $A_3 = 5$. Using these results, and the same arguments, it is easy to see that

$$A_4 = (J_3 \oplus 0_3) + (0_1 \oplus A_3)$$

is a CP matrix realization of S_6, rank $A_4 = 5$ and cp-rank $A_4 = 6$. $\square$

Combining Theorem 3.17 with Corollary 3.5 and Propositions 3.7 and 3.8 we obtain a characterization of the graphs which satisfy (3.31).

Theorem 3.18   *Let G be a connected graph, then the following are equivalent:*

(a) *For every CP matrix realization A of G, cp-rank $A = $ rank A.*
(b) *G contains no even cycle and no triangle free graph with more edges than vertices.*
(c) *Each block of G is a K_2 or an odd cycle, and at most one of the blocks of G has more than 3 vertices.*

Exercises

3.24   By Proposition 3.8 there exists a CP matrix realization of S_6 of rank 5 and cp-rank 6. Show that there exists also a CP matrix realization of S_6 with rank and cp-rank both equal to 5.

3.25   Show that if A is a CP matrix realization of S_6 and rank $A \neq 5$, then cp-rank $A = $ rank A.

Notes.   This section is based on [Shaked-Monderer (2001)].

Bibliography

T. Ando, Totally positive matrices, *Linear Algebra Appl.*, 90:165–219 (1987).

T. Ando, *Completely Positive Matrices*, Lecture Notes, The University of Wisconsin, Madison (1991).

R. B. Bapat, T. E. S. Raghavan, *Nonnegative Matrices and Applications*, Cambridge University Press, Cambridge (1997).

F. Barioli, Completely positive matrices with a book-graph, *Linear Algebra Appl.*, 277:11–31 (1998a).

F. Barioli, Decreasing diagonal elements in completely positive matrices, *Rend. Sem. Mat. Univ. Padova*, 100, 13–25 (1998b).

F. Barioli, Chains of dog-ears for completely positive matrices, *Linear Algebra Appl.*, 330:49–66 (2001a).

Duality and graph-theoretic tools for complete positivity of matrices, Ph.D. thesis, Padova university (2001b).

F. Barioli, Complete positivity of small and large accute sets of vectors, Talk at the 10th ILAS conference in Auburn.

F. Barioli, A. Berman, The maximal cp-rank of rank k completely positive matrices, *Linear Algebra Appl.*, 363:17–33 (2003).

W. Barrett, C. R. Johnson and R. Loewy, The real positive definite completion problem: Cycle completability, *Mem. Amer. Math. Soc.*, 584 (1996).

W. Barrett, C. R.Johnson and R. Loewy, Critical graphs for the positive definite completion problem, *SIAM J. Matrix Anal. Appl.*, 20:117–130 (1998).

F. A. Beardon, *Complex Analysis*, John Wiley & Sons, Chichester (1979).

F. Benatti, Floreanini, R. Romano, Complete positivity in dissapative quantum dynamics, *Dynamics of Dissipation*, P. Garbaczewski, R. Olkiewicz (Eds.), Lecture Notes in Physics, Springer-Verlag, Heidelberg 597:283–304 (2003).

A. Ben-Israel, T. N. E. Greville, *Generalized Inverses: Theory and Applications*, 2nd edition Springer-Verlag, New York (2003).

A. Berman, *Cones, Matrices and Mathematical Programing*, Lecture notes in economics and mathematical systems 79, Spinger-Verlag, Berlin (1973).

A. Berman, Complete positivity, *Linear Algebra Appl.*, 107:57–63 (1988).

A. Berman, Completely positive graphs, *Combinatorial and Graph-Theoretical Problems in Linear Algebra*, IMA Vol. Math. Appl., 50:229–233, Springer, New York (1993).

A. Berman, S. Gueron, On the inverse of the Hilbert matrix, *The Mathematical Gazette*, 86:274-277 (2002).

A. Berman, R. Grone, Completely positive bipartite matrices, *Math. Proc. Cambridge Philos. Soc.*, 103:269–276 (1988).

A. Berman, D. Hershkowitz, Combinatorial results on completely positive matrices, *Linear Algebra Appl.*, 95:111–125 (1987).

A. Berman, R. Plemmons, *Nonnegative Matrices in the Mathematical sciences.* SIAM, Philadelphia (1994).

A. Berman, N. Shaked-Monderer, Remarks on completely positive matrices, *Linear and Multilinear Algebra*, 44:149–163 (1998).

J. A. Bondy, U. S. R. Murty, *Graph Theory with Applications*, North-Holland, Amsterdam (1976).

S. Bose, E. Slud, Maximin efficiency-robust tests and some extensions, *J. Statist. Plann. Inference*, 46:105–121 (1995).

M. D.Choi, Tricks or treats with the Hilbert Matrix, *American Math. Monthly*, 90:301–312 (1983).

R. W. Cottle, Manifestations of the Schur complement, *Linear Algebra Appl.* 8:189–211(1971).

J. E. Cohen, U. G. Rothblum, Nonnegative ranks, decompositions, and factorizations of nonnegative matrices, *Linear Algebra Appl.*, 190:149–168 (1993).

D. E. Crabtree, E. V. Haynsworth, An identity for the Schur complement of a matrix, *Proc. Amer. Math. Soc.* 22: 364–366 (1969).

C. W. Cryer, Some properties of totally positive matrices, *Linear Algebra Appl.*, 15:1–25 (1976).

P. H. Diananda, On nonnegative forms in real variables some or all of which are nonnegative, *Proc. Cambridge Philos. Soc.*, 58:17-25 (1962).

J. H. Drew, C. R. Johnson, The no long odd cycle theorem for completely positive matrices. *Random Discrete Sturctures*, IMA Vol. Math. Appl., 76:103–115 (1996).

J. H. Drew, C. R. Johnson, The completely positive and doubly nonnegative completion problems, *Linear and Multilinear Algebra*, 44:85–92 (1998).

J. H. Drew, C. R. Johnson, S. J. Kilner and A. M. McKay, The cycle completable graphs for the completely positive and doubly nonnegative completion problems, *Linear Algebra Appl.*, 313:141–154 (2000).

J. H. Drew, C. R. Johnson and F. Lam, Completely positive matrices of special form, *Linear Algebra Appl.*, 327:121–130 (2001).

J. H. Drew, C. R. Johnson and R. Loewy, Completely positive matrices associated with M-matrices, *Linear and Multilinear Algebra*, 37:303–310 (1994).

C. H. Fitzgerald, R. A. Horn, On fractional Hadamard powers of positive definite matrices, *J. Math. Anal. Appl.* 61: 633-642 (1977).

F. R. Gantmacher, *The Theory of Matrices*, 2 vols. Chelsea, New York (1959).

F. R. Gantmacher, M. G. Krein, *Oscillation Matrices and Kernels and Small Vibrations of Mechanical Systems: Revised Edition*, AMS Chelsea Publishing (2002).

M. Gasca, C. A. Michelli (eds.), *Total Positivity and its Applications*, Kluwer Academic Publishers, Dordrecht (1996).

F. Goldberg, On completely positive graphs and their complements, *Linear Algebra Appl.*, (2003).

M. Golumbic, *Algorithmic Graph Theory and Perfect Graphs*, Academic Press, New york (1980).

L. J. Gray, D. G. Wilson, Nonnegative factorization of positive semidefinite nonnegative matrices. *Linear Algebra Appl.*, 31:119–127 (1980).

R. Grone, C. R. Johnson, E. M. Sá and H. Wolkowitz, Positive definite completions of partial hermitian matrices, *Linear Algebra Appl.*, 58:109–124 (1984).

M. Hall Jr., A survey of combinatorial analysis, *Some aspects of analysis and probability*, Surveys in Applied Mathematics IV, John Wiley & Sons, 35–104 (1958).

M. Hall Jr., *Combinatorial Theory*, 2nd edition, John Wiley & Sons, (1986).

M. Hall Jr., M. Newman, Copositive and completely positive quadratic forms, *Proc. Cambridge Philos. Soc.*, 59:329-339 (1963).

C. L. Hamilton-Jester, C-K. Li, Extreme Vectors of Doubly Nonnegative Matrices, *Rocky Mountain J. Math.*, 26:1371-1383 (1996).

J. Hannah, T. J. Laffey, Nonnegative factorization of completely positive matrices. *Linear Algebra Appl.*, 55:1–9 (1983).

E. V. Haynsworth, Determination of the inertia of a partitioned Hermitian matrix, *Linear Algebra Appl.*, 1:73–81(1968).

R. A. Horn, On infinitely divisible matrices, kernels and functions, *Z. Warsh. Verw. Gebete* 8:219-230 (1967).

R. A. Horn, The theory of infinitely divisible matrices and kernels, *Trans. Amer. Math. Soc.* 136:269-286 (1969).

R. A. Horn, C. R. Johnson, *Matrix Analysis*, Cambridge University Press, Cambridge (1985).

R. A. Horn, C. R. Johnson, *Topics in Matrix Analysis*, Cambridge University Press, Cambridge (1991).

D. H. Jacobson, Factorization of symmetric M-matrices, *Linear Algebra Appl.*, 9:275-278 (1974).

C. R. Johnson, R. L. Smith, The completion problem for M-matrices and inverse M-matrices, *Linear Algebra Appl.*, 241–243:655-667 (1996).

S. Karlin, *Total Positivity*, Vol. 1, Stanford University Press, Stanford (1968).

M. Kaykobad, On nonnegative factorization of matrices *Linear Algebra Appl.*, 96:27-33 (1987).

C. Kelly, A test of the Markovian model of DNA evolution, *Biometrics*, 50: 653-664 (1994).

Completely positive graphs and completely positive matrices, M.Sc. Thesis (in Hebrew), Technion (1989).

N. Kogan, A. Berman, Characterization of completely positive graphs, *Discrete Math.*, 114:298–304 (1993).

M. G. Krein, M. A. Rutman, Linear operators leaving invariant a cone in a Banach space, *Uspehi Math. Nauk (N.S.) 3*, 23:3–95(1948). (English translation in *Amer. Math. Soc. Transl. Ser. 1*, 10:199–325, Providence, R.I. (1956).)

C. M. Lau, T. L. Markham, Square triangular factorizations of completely positive matrices, *Indust. Math.*, 28:15–24 (1978).

R. Loewy, B-S. Tam, CP rank of completely positive matrices of order 5, *Linear Algebra Appl.*, 363:161–176 (2003).

M. Marcus, H. Minc, Some results on doubly stochastic matrices, *Proc. Amer. Math. Soc.* 13:571–579 (1962).

M. Marcus, H. Minc, *A Survey of Matrix Theory and Matrix Inequalities*, Allyn and Bacon, Boston (1964).

T. L. Markham, Factorization of completely positive matrices, *Proc. Cambridge Philos. Soc.* 69:53–58 (1971).

J. E. Maxfield, H. Minc, On the matrix equation $X'X = A$, *Proc. Edinburgh Math. Soc.* 13(II):125–129 (1962).

H. Minc, *Nonnegative Matrices*, John Wiley & Sons, New York (1988).

T. Motzkin, Copositive quadratic forms, *National Bureau of Standards Report* 1818:11–12 (1952).

R. T. Rockafellar, *Convex Analysis*, Princeton University Press, Princeton (1970).

L. Salce, *Lezioni Sulle Matrici*, Decibel-Zanichelli, Padova (1993).

L. Salce, P. Zanardo, Completely positive matrices and positivity of least sqaure solotions, *Linear Algebra Appl.*, 178:201–216 (1993).

I. Schur, Potenzrihen im Innern des Einehitskreiss, *J. Reine Angew. Math.*, 147:205–232 (1917).

E. Seneta, *Nonnegative Matrices: An Introduction to Theory and Applications*, Allen and Unwin, London (1973).

N. Shaked-Monderer, Minimal cp rank, *Electron. J. Linear Algebra*, 8:140–157 (2001).

L. E. Trotter, Jr. Line perfect graphs, *Math. Program.*, 12:255–259 (1977).

R. S. Varga, *Matrix Iterative Analysis*, 2nd Edition, Springer-Verlag, Berlin (2000).

W. Watkins, The cone of positive generalized matrix functions, *Linear Algebra Appl.*, 181:1–28 (1993).

S. Xiang, S. Xiang, Notes on completely positive matrices, *Linear Algebra Appl.*, 271:273–282 (1998).

C-Q. Xu, Completely positive matrices of order five, *Acta Math. Sinica, Enlgish series*, 17:550–562 (2001)

C-Q. Xu, Nearly completely positive graphs, *Appl. Algebra Engrg. Comm. Comput.*, 13:1–8 (2002).

C-Q. Xu, J-S. Li, A note on completely positive graphs, *Syst. Science and Math.*

Sciences, 13:121–125 (2000).

B. Ycart, Extrémales du cône des matrices de type non négatif, à coefficients positifs ou nals, *Linear Algebra Appl.*, 48:317–330 (1982).

X-D. Zhang, J-S. Li, Completely positive matrices having cyclic graphs, *J. Math. Res. Exposition*, 20:27–31 (2000).

X-D. Zhang, J-S. Li, Factorization index sets for completely positive graphs, *Acta Math. Sinica, Enlgish series*, 18:823–832 (2002).

Notation

Most of the symbols are listed together with a brief description and a page reference to the first appearance. General notations appear with no page reference. Some ad hoc notations appear with page reference only.

$A(\alpha|\beta]$	submatrix of A, rows indexed by $\{1,\ldots,n\}\setminus\alpha$ and columns by β	
$A(\alpha|\beta)$	submatrix of A, rows indexed by $\{1,\ldots,n\}\setminus\alpha$ and columns by $\{1,\ldots,n\}\setminus\beta$	
$A \circ B$	Hadmard (Schur) product of A and B	13
$A \oplus B$	direct sum of A and B	3
$A \otimes B$	Kronecker product of A and B	14
$A \geq B$	$A - B$ is (entrywise) nonnegative	18
$A \succeq B$	$A - B$ is positive semidefinite	10
adj A	classical adjoint of A	
A/E	(genrealized) Schur complement of a E in A	24,27
$A/[i]$	$A/A[i]$	25
$A^{(k)}$	k-th Hadamard power of A	14
$A_{(i)}$	i-minimized matrix of A	67
$A^{[r]}$		119
A^R		111
A^S		111
A^T	transpose of A	
C	the complex numbers	
$c(G)$	edge-clique cover number of G	144
cl K	closure of K	43
C_n	n-cycle	32
cone S	convex cone generated by S	42
$\mathcal{COP}_n$	cone of $n \times n$ copositive matrices	51
conv S	convex set generated by S	42
$\mathcal{CP}_n$	cone of $n \times n$ completely positive matrices	71
cp-rank A	cp-rank of A	139
cp-rank G	cp-rank of G	145
cs A	column space of A	
$\mathcal{C}_W$	cone generated by the columns of W	147
$D(A)$	directed graph of A	38
$\delta(A)$		112
$\Delta_A(x)$	characteristic polynomial of A	6
det A	determinant of A	
$d_G(u,v)$	distance in G between u and v	32
diag($\mathbf{x}$)	diagonal matrix with the entries of $\mathbf{x}$ on the main diagonal	2

$\mathcal{DNN}_n$	cone of $n \times n$ doubly nonnegative matrices	50
$d(v)$	degree of v	31
$\mathbf{e}$	column vector of all ones	2
$E(G)$	set of edges of G	30
$\mathbf{e}_i$	i-th standard basis vector	2
E_{ij}	matrix with all zero entries except for 1 in the ij position	2
$f(A)$	function of A	130
$f_H(A)$	Hadamard function of A	130
$\overline{G}$	complement of G	31
$G(A)$	graph of A	39
$G + e$	graph obtained by adding edge e to G	32
$G - e$	graph obtained by removing edge e from G	32
$G_1 \cap G_2$	intersection of G_1 and G_2	32
$G_1 \cup G_2$	union of G_1 and G_2	32
$\mathrm{Gram}(\mathbf{v}_1, \ldots, \mathbf{v}_n)$	Gram matrix of $\mathbf{v}_1, \ldots, \mathbf{v}_n$	11
$G - v$	subgraph induced by vertices other than v	32
$H \subseteq G$	H is a subgraph of G	31
$H(n)$	$n \times n$ Hilbert matrix	5
$i(A)$	inertia of A	7
$i_+(A)$	number of positive eigenvalues of A	7
$i_-(A)$	number of negative eigenvalues of A	7
$i_0(A)$	multiplicity of 0 as an eigenvalue of A	7
I_n	identity $n \times n$ matrix	2
$\mathrm{int}\, K$	interior of K	43
$J_{m \times n}$	$m \times n$ matrix of all ones	2
J_n	$J_{n \times n}$	2
$K_1 + K_2$	sum of K_1 and K_2	43
$K_{m,n}$	complete bipartite graph	34
K_n	complete graph on n vertices	31
$K_{\mathbf{u}, \theta}$	circular cone	42
$L^{\perp}$	orthogonal complement	
$L(G)$	Laplacian matrix of G	37
$M(A)$	comparison matrix of A	3
$N(G)$	adjacency matrix of G	37
$\mathcal{NN}_n$	cone of $n \times n$ nonnegative matrices	49
$\mathrm{ns}\, A$	nullspace of A	

Index